Bluetooth® LE Audio

This book unravels the technical intricacies and recent developments of Bluetooth LE Audio technology. Covering everything from the radio frequency model and LC3 Codec to application profiles and use cases, it provides a deep dive into the new architecture that is transforming wireless audio. It offers clear insights into the transition from Bluetooth® Classic to LE Audio, emphasizing power efficiency, new features like Auracast™ broadcast audio, and the innovative potential of LE Audio in various devices.

- Provides a detailed look into all aspects of LE Audio, covering radio frequency models, logical links, transport layers, and the advantages of the LC3 Codec
- Explains the shift from Classic Audio to LE Audio, highlighting the advantages in power savings, efficiency, and new use cases enabled by Bluetooth LE technology
- Breaks down the architecture of the LE Audio stack, from the Core Controller to Host protocols, profiles, and application interfaces, making it easier for developers to navigate the standard
- Explores cutting-edge features like Auracast broadcast audio for broadcasting audio content and synchronized audio streaming for multi-device setups, offering insights into future innovations in audio technology
- Emphasizes the potential for significant power savings with LE Audio, especially for devices with smaller batteries, allowing for longer usage without frequent charging

This book is a must-read for developers and engineers seeking to master the latest advancements in Bluetooth technology.

Bluetooth® LE Audio

Fundamental to Recent Advances

Himanshu Bhalla and Oren Haggai

CRC Press
Taylor & Francis Group
Boca Raton London New York

CRC Press is an imprint of the
Taylor & Francis Group, an **informa** business
A CHAPMAN & HALL BOOK

First edition published 2026
by CRC Press
2385 NW Executive Center Drive, Suite 320, Boca Raton FL 33431

and by CRC Press
4 Park Square, Milton Park, Abingdon, Oxon, OX14 4RN

CRC Press is an imprint of Taylor & Francis Group, LLC

ISBN: 978-1-032-96619-9 (hbk)
ISBN: 978-1-032-96621-2 (pbk)
ISBN: 978-1-003-59018-7 (ebk)

DOI: 10.1201/9781003590187

Typeset in Times
by Apex CoVantage, LLC

Contents

Preface

In this book, we tell you the story behind the evolution of the LE Audio technology.

In the first three chapters, we will review audio in general, Bluetooth® in general, and the high-level architecture of LE Audio. The concepts that we will share with you in the first three chapters may sound familiar in some parts and may sound new in other parts. The purpose of the first three chapters is to focus on the motivation for LE Audio and what experience LE Audio is creating for users. This is the promise of LE Audio, which we hope to deliver to you.

The following four chapters will go into more depth on the LE Audio concepts. You will learn more details about the various specifications of LE Audio and how these specifications apply to the LE Audio Architecture. We will review the protocols each of the specifications follows in order to deliver the LE Audio promise for new use cases.

In the last chapter, we will take you back to the basics. We will illustrate to you use case examples from an application point of view and how the LE Audio orchestra of specifications is playing together to deliver a new type of interoperable user experience.

We hope you will find this book useful, and we thank you for taking the time to read it. It summarizes the efforts of over five years of specification development done by multiple Bluetooth Special Interest Group (SIG) Working Groups, with over 50 individual contributing members from over 25 different member companies.

The fusion between Bluetooth® technology and audio provides the ultimate personal experience to people. It is also the fusion between acoustic signal processing and Radio frequency signal processing.

Acknowledgments

HIMANSHU

Thank you, my wife, Esha, for helping me find my happiness in this work and nudging me to continue working on Bluetooth® technology at a critical professional juncture in life. There were countless hours of family and holiday time that I spent writing this book. You always greeted me with a smile and affection after I had finished writing, which gave me the power to continue working on.

Thanks to our daughters, Alaina and Azaira – I wish that someday you would read this book.

Thank you, Mom and Dad – you have always inspired me to keep moving forward and break all the barriers. I am what I am because of all the values and learnings that you provided. I will forever be grateful.

Thank you, Oren, for agreeing to work on this book ("Wild idea"). I still remember you showing confidence in me and handing over the editorship of the critical ASCP (now split into BAP, ASCS, and PACS) specification. That was the turning point that led to the foundations of this book. You have been a mentor to me in many aspects, and I sincerely thank you for that.

Thank you to all the fellow engineers who have worked tirelessly over the past so many years in creating, developing, and innovating Bluetooth Technology and progressing Humanity forward in our own small way.

OREN

First and foremost, I would like to thank my wife, Limor, for being so loving and caring, for making my life meaningful each day, and for motivating me while working on this book. Limor, over the years, you are the light in my journey, giving me great advice when I most need it. Your wisdom guides and helps me grow to become a better person. 28 years ago, when I was a principal electrician working in stone quarries and asphalt plants, it was you who guided me to make a career shift, go to university, and get a BSc degree in EE and computer science. Thanks to our dear daughter, Lori, for her love and patience over the years. Lori, it is my sheer pleasure to watch you grow, get wiser and contributing so much to yourself and the people around you, I learn from you a lot and admire everything you do.

Thanks to my colleagues throughout the years, for the time we spent together providing better wireless solutions. Thanks to the people in the Bluetooth® SIG core and audio working groups for keeping up and striving to define specifications that improve the Bluetooth audio ecosystem. And thank you especially to Himanshu for initiating the idea to write a book about LE Audio. It was my sheer pleasure to work with him on the book, and our collaboration proved to be fruitful.

About the Authors

Himanshu Bhalla is a seasoned Software Architect at Intel Corporation's Client Computing Group in Bengaluru, India. A graduate in Computer Science & Engineering from IIT-BHU, he boasts nearly 20 years of expertise in software architecture, design, and development, focusing on Bluetooth® and wireless technologies. He began at a startup, Impulsesoft, a pioneer in Bluetooth-based audio headsets, then moved to Broadcom as Bluetooth technology architect for Samsung Tizen DTV. His contributions to Wi-Fi Direct Services (WFDS) and Wi-Fi Alliance are notable. At Intel, he shaped the Wi-Fi Serial Bus (WSB) standard and pioneered MAUSB (Media Agnostic USB) technology. Bhalla played a pivotal role in defining the LE Audio standard, earning "Outstanding New Contributor of the Year" from Bluetooth SIG in 2018. He led the Bluetooth SIG Internet Working Group, advised the Bluetooth Architecture Review Board (BARB), and critiqued 50+ Bluetooth SIG specifications. An inventor with multiple USPTO patents, Bhalla credits his work as a tribute to technology and its community.

Oren Haggai is a wireless systems expert at Apple, specializing in Bluetooth® technology with over 25 years of expertise. He began his journey with Bluetooth technology in 1999 while studying at Technion University, integrating Bluetooth stacks for handheld PCs and phones. At Mobilian, he contributed to the first Wi-Fi and Bluetooth technology-based coexistence solution and designed the inaugural wireless desktop system coexisting with Wi-Fi networks. Joining Intel in 2003 post-Mobilian's acquisition, he developed Bluetooth technology-based solutions for phones and PCs, crafting quality of service-based cellular WiMAX network connection managers. Haggai trained diverse Intel teams in Bluetooth technology and authored system requirements for Bluetooth® LE products. He played a pivotal role in shaping LE Audio specifications and holds patents in wireless communications. Active in Bluetooth SIG Working Groups, he contributed significantly to LE Audio Controller, Host, and Codec aspects, pioneering LE Audio terms like PAC (Published Audio Capability) and BASE (Broadcast Audio Source Endpoint). Haggai is passionate about advancing the evolution of Bluetooth technology and sharing this excitement with a broader audience.

1 Introduction

LE Audio is a large set of new specifications, which were developed by the Core and the Audio Working Groups in the Bluetooth® SIG. With LE Audio, Bluetooth technology is addressing audio from a different perspective than the previous generation of audio over Bluetooth. The previous generation of Bluetooth audio is known as Classic Audio.

In Classic Audio, different audio use cases were addressed by different methods and defined in separate specifications. LE Audio uses generic methods to address all use cases. LE Audio grasps audio as a whole, addressing all existing use cases but without limiting the technology to a specific use case.

With LE Audio, use cases that were not possible with Classic Audio are now becoming possible. Because LE Audio is addressing audio using a uniform layered approach, future use cases can use the same architecture framework to build upon.

Many existing use cases will consume substantially less power as the radio duty cycle of audio over Bluetooth® LE is lower when compared with Bluetooth® Classic.

In this chapter, we will focus on the motivation behind LE Audio while providing a brief about the architecture. This chapter will also review the requirements from the perspective of various audio use cases and how these were addressed by Classic Audio and compare it to LE Audio.

MOTIVATION FOR LE AUDIO

LE Audio is designed to address a wide range of use cases and configurations. It is built on top of Bluetooth® LE, which provides media access radio technology with advantages over Bluetooth Classic. The main advantages of Bluetooth® LE are faster connection time, lower duty cycle in steady-state connection, and predictive scheduling of traffic. Bluetooth LE technology is proven to be more efficient in power consumption due to the preceding three reasons. Peak power is often attributed to longer connection time. Peak power also correlates to transitions between low-power states and active power states. In Bluetooth LE, the connection time and transition to low-power state are faster, leading to lower power consumption, compared to Classic Audio.

These improvements enable use cases in LE Audio for all-day usage on form factors that cannot have large batteries, like hearing aid devices. Hearing aid users need to use the hearing aid devices for a longer duration of time without the need to charge often. The hearing aid users expect their devices to connect to various audio sources, and therefore, a short connection time provided by Bluetooth LE is essential.

The scheduling of traffic over Bluetooth LE is done at a deterministic rate, which allows scheduling of traffic with lower energy consumption due to better planning and bandwidth allocation. Another advantage of deterministic rate scheduling is the ability to schedule multiple connections and streams to a set of devices with minimal

DOI: 10.1201/9781003590187-1

conflicts or collisions. This allows power-efficient sharing of radio spectrum among multiple audio devices consuming the same audio content, for example, left and right earbud use case – where the same audio content is streamed to two independent earbuds. And it allows power-efficient sharing of radio spectrum when multiple streams with different contexts are multiplexed on the same physical transport – for example, gaming use cases where music and voice are streamed together.

An essential aspect of Bluetooth LE is the Broadcast capability. With Bluetooth LE, application data may be broadcast to an unlimited number of listeners using the concept of Advertising. Expanding this capability to LE Audio means that audio content can be published and shared across an unlimited number of listeners – bringing interesting Broadcast use cases for the first time in Bluetooth technology-based audio. Broadcast audio in LE Audio is also known as Auracast™ broadcast audio. This differentiates LE Audio Broadcast from other forms of Broadcast in Bluetooth LE, such as the various Advertising transports (we will discuss these aspects in more detail in later chapters).

LE Audio is extensible, and it is forward compatible, which means that it provides frameworks that facilitate the development of new use cases. LE Audio provides flexibility in the communication of audio content for a use case. At the same time, it also defines clear rules for interoperability. The result is a standard and interoperable ecosystem in which new use cases become possible between different classes of devices from various manufacturers.

Figure 1.1 shows a high-level audio stack comparison between Classic Audio and LE Audio. In Classic Audio, the audio stack has different components for handling voice and music/media audio. In LE Audio, the same set of protocols handles both voice and music/media audio. In Classic Audio, the topology of audio connections over the radio link is based on point-to-point, single connections. In LE Audio, coordination and synchronization between multiple devices are possible when audio content is sent over the radio. In Classic Audio, different sets of audio codecs are used for voice or music/media audio. In LE Audio, a single mandatory Codec is used

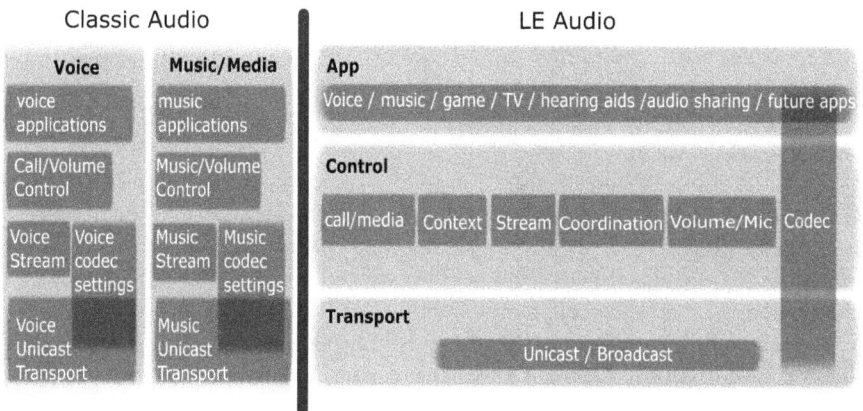

FIGURE 1.1   Evolution from Classic Audio to LE Audio.

for both voice and music/media audio and applications, while optional and vendor-specific codecs can also be provisioned to extend the technology.

LE AUDIO ARCHITECTURE

Table 1.1 describes the layers of the LE Audio Architecture. Each horizontal layer in the table is a set of specifications, which may be protocols, profiles, or services. The vertical Codec layer has an impact on all the horizontal layers. In later chapters in this book, we will review each layer in more depth. In this section, we provide a general overview of each layer. In this book, we will introduce you to a new concept which we call "The One Architecture." This is the author's view on the LE Audio Architecture. We will get back to this section at the end of this chapter and later at the beginning of Chapter 3.

TABLE 1.1

LE Audio Architecture Layers (The One Architecture)

Layer	Purpose	Description	
App	Targets end user	Interface to audio applications	
Control	Generic controls and audio data interfaces	Sets up audio content	Codec
Transport	Over-the-air audio	Scheduling of Unicast and Auracast™ Broadcast Audio Streams, which carry compressed audio frames	

App Layer

The App layer defines common methods and interfaces used by various types of audio applications that use LE Audio. It specifies the set of selected generic components from the Control layer in order to realize the use cases. The App layer configures the Control layer with audio settings and Codec settings for the desired audio quality of the use case. Example applications are high-quality media playback, TV Broadcast, hearing aids, gaming consoles, public announcement systems, and so on. The use of the Control layer allows extensibility for future applications, which can use different configurations and different combinations of controls for fulfilling the requirements of a use case.

Control Layer

The Control layer provides a rich set of generic controls that address all aspects of wireless audio. Table 1.2 lists the various controls that are available in LE Audio. Each control is self-contained and serves a unique function. This approach allows extending the LE Audio Architecture in the future without affecting existing functionality. This is a powerful feature of LE Audio technology that enables both backward compatibility and forward compatibility.

TABLE 1.2

LE Audio Control Layer Blocks

Control Layer Component	Purpose	Description
Stream	Sets up Audio Streams	Discovers audio Codec capabilities and creates coded Audio Streams for Unicast audio or Auracast™ broadcast audio
Call	Manages call control	Answers or makes calls
Media	Manages playback control	Plays and controls media tracks
Context	Binds context to peripherals	Controls assignment of Audio Context in a use case, uniformly across all participating audio peripherals
Volume	Manages volume	Increases, decreases, or mutes the volume level of an audio peripheral
Microphone	Manages microphone	Activates or mutes a microphone on an audio peripheral
Coordination	Connects to a set of peripherals	Discovers, authenticates, and connects to a set of audio peripherals, as a single unit

Within the Control layer, each block serves a different function. The functionality of each block was carefully selected to avoid overlap.

The Stream control deals with the discovery, configuration, and setup of Audio Streams. Discovery is the act of probing the remote audio peripheral capabilities – including compression and decompression capabilities. By discovering what compression types the remote device supports, the stream configuration may be tuned to support the required audio settings per use case. The App layer uses the Stream control to configure the audio settings of the stream. Stream control allows the App layer to control the enabling or disabling of the Audio Streams.

The App layer may use different types of streams, which are mainly divided into Unicast streams or Broadcast streams. The App layer may use a single stream to mix different audio use cases or context types. For example, the same stream may carry either call audio or music/media audio. The App layer may apply the same Audio Context and stream settings across multiple audio peripherals. Applying the Audio Context and stream settings across multiple peripherals is done via the Context control procedures.

The Call and Media are context type controls, which provide functionality to control calls and media playback, respectively. Each of these remote controls is considered a different context type that the Context layer may deploy when setting multiple streams across one or more audio peripherals. The App layer sets up a stream per use case need and follows the common procedures that are defined by the Context control.

The Context control provides the common procedures for the App layer to follow when use case content is deployed over streams and multiple audio peripherals. The Context control defines procedures for starting and ending audio over sets of audio peripherals, and other procedures to update the context of streaming audio and control speakers and microphone gain on the set of peripherals. The Context control defines procedures to enable transmission of different types of contexts to a single device or to multiple devices and how the multiple devices are synchronized to the use case. These procedures are extensible, and any future App layer profile may use these generic procedures while achieving basic interoperability.

The Volume and Microphone controls provide a unified method to control the volume and gain on a speaker or a microphone within a single audio peripheral. The Context control synchronizes multiple peripherals when the speakers and microphones span across multiple devices that are playing the same use case. Using a unified control of volume over a device enhances the user experience. The user may locally control the volume or remotely control the volume regardless of the use case or context type.

Coordination controls how to discover, authenticate, and connect to a set of peripheral devices. This control allows multiple use cases to use the same set of devices by controlling when the set of devices is currently in use. While a set is in use, it may not be used by a different remote device.

Transport Layer

The Transport layer defines over-the-air transport and its parameters to support LE Audio Streams as required by the Control layer. Two types of transport are defined to support Unicast audio and Broadcast audio. Both of these transports provide a fixed interval and window and a rich set of parameters to control the quality of service. The quality of service parameters enables reliability, latency, and required bitrate. The Control layer uses the quality of service parameters to select the Transport layer configuration which would ultimately satisfy the required Codec setting as per the App layer needs for a given use case.

In the case of Broadcast, the transport is unidirectional, and packets are sent with multiple copies as per the reliability configuration. Broadcast is connectionless, and no feedback is possible. Sending multiple copies of the same content increases the chance of listeners receiving a correct copy of the data. The receivers filter out redundant Broadcasted copies before forwarding the audio data for application playback.

In the case of Unicast, the transport is bidirectional. There are two reasons for that. The first reason is to enable a reliable feedback mechanism for acknowledging the reception of the audio packets. The second reason is allowing the flow of audio in two directions, to generate and consume audio as part of a typical audio use case – like an audio call where both speaker and microphone are active at the same time and audio flows in both directions.

Figure 1.2 illustrates the concept of Broadcast transport and Unicast transport. A single Unicast transport is serving a single user, while a single Broadcast transport is serving many users.

While using Broadcast transport, compressed audio packets are sent in one direction and in multiple copies (Figure 1.2 shows three copies). Multiple users may synchronize to the Broadcast stream and receive the audio packets while filtering out redundant copies. Broadcast streams may carry voice audio or music/media audio. Auracast™ broadcast audio applications may include broadcasting of terminal flight announcements in an airport, TVs in a sports bar, or an auditorium amplification system to share the live lecture to local loudspeakers and remote listeners over Bluetooth (the local loudspeakers may receive the audio wirelessly over Auracast broadcast audio).

While using Unicast transport, the stream may be used to carry media or music in one direction to a remote speaker or to carry a two-way voice communication. In any case, the over-the-air transport for Unicast audio is bidirectional to carry the acknowledgment of packets in addition to audio in the reverse direction (if any). The smaller packets in Figure 1.2 represent acknowledgment. Unicast transport takes turns to send packets in each direction using a time division duplex mechanism, which is further described in Chapter 2 (Bluetooth® LE Overview).

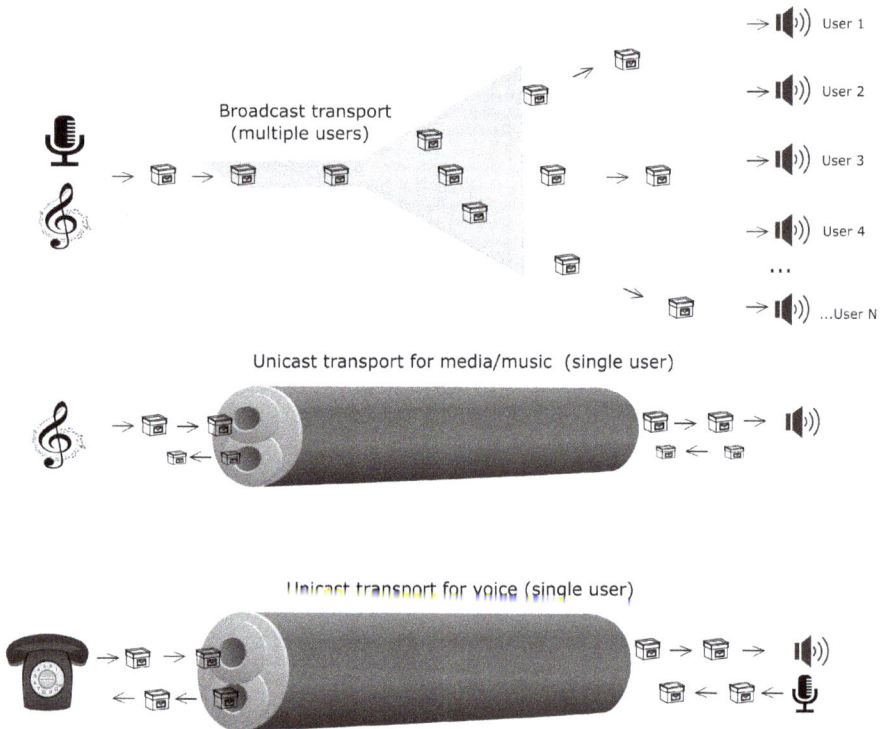

FIGURE 1.2   Auracast™ transport and Unicast transport.

Figure 1.3 shows how Broadcast and Unicast transports in LE Audio may scale up to carry a group of transports. In the Auracast broadcast audio case, the example is Broadcast audio from a TV set in a public restaurant. Multiple users may synchronize to the Broadcast stream and listen to the TV using wireless LE Audio earbuds or hearing aids. The TV does not use its local speakers. Instead, the TV transmits audio over Auracast broadcast audio. The Broadcast transport in this case contains a group of streams. In this example, two streams are shown to carry audio in two languages, such as English dubbing and Spanish dubbing. The user may select which stream to synchronize to, based on their language of choice.

In the Unicast case, an example of a multichannel surround system is shown. The stereo system transmits music to four devices: surround left, sound bar, subwoofer, and surround right. The sound bar contains center, left, and right channels. The stereo system may also receive phone calls, and a microphone attached to the surround left speaker captures voice audio from the room. This example illustrates three concepts. The first concept is that multiple audio channels representing multiple locations may be multiplexed into a single stream and a single transport (the sound bar in this example). The second concept is that a given audio transport may allow multiple contexts, such as voice and music (the surround left and microphone in this example). The third concept is that a collection of devices may be connected as a single set of coordinated devices for a given use case and form a group of transports that are serving a single use case (the entire surround system in this example).

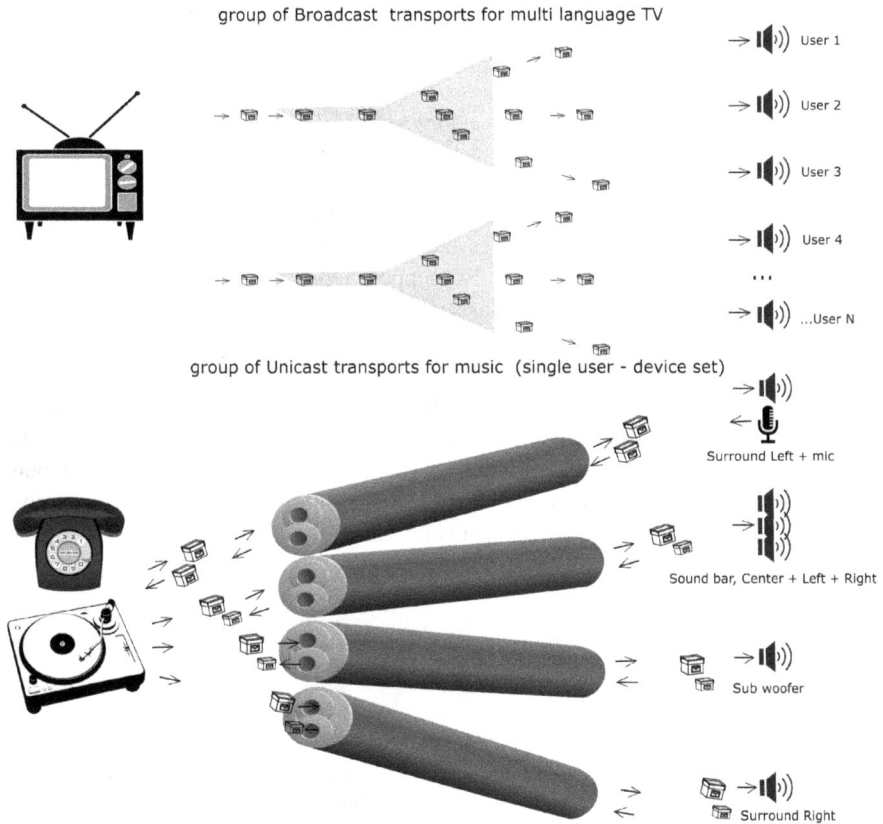

group of Broadcast transports for multi language TV

User 1
User 2
User 3
User 4
...
...User N

group of Unicast transports for music (single user - device set)

Surround Left + mic

Sound bar, Center + Left + Right

Sub woofer

Surround Right

FIGURE 1.3 Group of transports for Broadcast or Unicast use cases.

CODEC

As seen in both Figure 1.1 and Table 1.1, the Codec layer spans across the App, Control, and Transport layers and provides compression and decompression of audio frames. It is tightly coupled with the data path of audio and, therefore, described as a separate vertical layer.

LE Audio supports a default mandatory Codec with a wide range of sampling rates and compression options. LE Audio allows the usage of an optional Codec (externally defined) or a vendor-specific Codec. All devices supporting LE Audio must support the default mandatory Codec and may support other optional codecs or vendor-specific codecs. All audio traffic in LE Audio is compressed before sending it over the air and decompressed after receiving it from over the air. Compression reduces the amount of radio energy required and lowers the duty cycle during which the Bluetooth® radio must remain active. Reducing the duty cycle frees more spectrum to deploy audio with multi-device and multi-location use cases.

The default mandatory Codec in LE Audio is achieving a high compression ratio while maintaining an excellent audio quality. The compression factor is roughly eight. We will review the mandatory Codec in more detail in later chapters.

COMPATIBILITY WITH CLASSIC AUDIO

Devices may support LE Audio only, support both Classic Audio and LE Audio, or support Classic Audio only. Table 1.3 shows the compatibility matrix of possible support combinations between two devices. Older devices that only support Classic Audio cannot work with devices that support LE Audio only. This is why a category called Dual Audio also exists. Devices such as PCs or phones support Dual Audio, while some audio peripherals may also support Dual Audio. These devices may work with both classes of devices: Classic Audio and LE Audio.

The case where LE Audio is in use is if both devices support LE Audio or Dual Audio. When the technology evolves, most devices use the Dual Audio class, since it allows wider compatibility. Over time, the number of LE Audio-only devices will grow. The Dual Audio class of devices is therefore essential during the transition of the Bluetooth® ecosystem toward LE Audio. This class is also what makes LE Audio both backward compatible and forward compatible. Over time, new use cases become possible only over LE Audio, and devices with only Classic Audio will become deprecated.

TABLE 1.3
Compatibility Matrix

\ Device A
   \
     \
 Device B \
	Support Only
	 LE Audio
		Support
		 Dual Audio
			Support Only
			 Classic Audio

Support Only **LE Audio**	Use LE Audio	Use LE Audio	Incompatible
Support **Dual Audio**	Use LE Audio	Use LE Audio	Use Classic Audio
Support Only **Classic Audio**	Incompatible	Use Classic Audio	Use Classic Audio

Before we continue and elaborate on the protocol aspects of LE Audio, we will take a detour. We will now review with you audio and digital audio in general, and what audio means to the end users like us.

AUDIO

Digital Audio provides a personal experience to people. People use audio to communicate with each other over telephone lines or voice over IP applications. People use audio to listen, share, or record their favorite music. Audio is unique in the sense that it enables people to multitask while listening to music or having a phone call with a friend.

The term digital audio as opposed to analog audio refers to how audio is communicated from an origin device to a destination device. The original audio content may still be analog. For example, a human voice is considered analog audio, since it is generated due to vibrations of the vocal cords, which in turn create acoustic waves that travel in the atmosphere as sound waves. In this case, the result is a continuous sound wave. The sound waves of the human voice consist of multiple tones or frequencies that are mostly concentrated in frequencies between around 300 Hz and around 8000 Hz, with certain tones which may exceed these limits. An opera singer may be able to generate such tones, which are higher than 8000 Hz.

A voice consists of a superposition of various tones and generates continuous signals in space time, with varying amplitudes. Other audio sources may be from musical instruments or sounds from nature (such as birds' sounds). The general sound spectrum may have very high tones. However, our ear is usually limited in sensitivity to what spectrum it may sense and send to our brain to process. The human ear may sense audio spectrum tones in a range of up to around 22,000 Hz. Some humans may have a more sensitive ear than others, but it is common to use the 22,000 Hz figure as the rough limit. The sound wave, therefore, consists of elements with changing frequencies and amplitude.

Digital audio allows us to represent the analog waves using computer representation, which means that a continuous wave is transformed into a variable number that changes over time. This conversion is known as analog to digital conversion. It is also known as Pulse Code Modulation (PCM). With PCM, the analog audio is sampled by filtering with very short pulses, thousands of times per second. According to the Shannon sampling theory, the sampling rate needs to be at least twice the spectrum content. Therefore, for voice content of 8000 Hz, a good PCM sampling rate is 16,000 Hz. And for music, a good sampling rate is 48,000 Hz, which is more than twice of what our ear sensitivity may consume (22,000 Hz).

Each sample provides a number that represents a point on the curved waveform. The waveform amplitude of the tone is represented by the maximum and minimum numbers, and it also expresses the volume level of the signal. Waveform amplitudes of the tones are often converted to units of dB to represent logarithmic amplification. As an example, the number range may be 16 bits signed in resolution, providing a range of −32,768 to +32,767. Sampling results in a large set of these numbers and adds up to a sample buffer. When the buffer needs to be transmitted over a medium, it often needs to be compressed, because otherwise the bitrate may be too high.

Compressing audio sample buffers may reduce their quality since some of the audio content may be lost. An example of a simple conversion from amplitude to a bit is known as waveform coding, or bit coding. This type of coding tries to track the waveform by indicating a plus or minus, using a single bit for polarity, 0 for minus, and 1 for plus. The polarity bit simply tracks the waveform derivative, whether it is going up or going down. Waveform coding is usually able to converge and follow the original signal. However, the response time in converging leads to the loss of quality and deviations, which cause unwanted noise artifacts.

A more sophisticated coding compression scheme may use a frequency domain transformation, by converting the time domain waveform to frequency domain tones. These conversions allocate a number of bits to encode different frequency bands. A band with more audio content gets more bits. There are other types of conversions, which are a combination of frequency and amplitude based on a model of human voice or sound content. The modern codecs are able to create better conversions by making certain assumptions about the sound content and using transformations that are more adaptive to the changing signal and its content.

Figure 1.4 shows a representation of a sound wave, as it changes in time. The height of the signal is the amplitude, which results in higher or lower gain levels of signal. This is how loud or muted the sound is. The tone is represented by the frequency of the signal, and it is how fast the signal changes over time. Digital audio is transforming the signal into a series of numbers, before compression.

FIGURE 1.4   Continuous sound wave.

Digital Audio is an extension of the human hearing sense. With digital audio, the hearing sense can be amplified to improve the hearing quality in terms of the hearing gain level and audio spectral content. For example, headphones provide the ability to increase or decrease the played audio volume; and in some situations, a different audio content is sent to the right and the left ears to enable what is known as the stereo effect.

Digital Audio is an extension of the human voice where language and speech can be either amplified or synthesized to create a new form of communication between humans or between a human and a machine. For example, with speech amplification (via a microphone), a single person may communicate to a large number of people in a theater without having to shout and damage the vocal cords. When voice is digitized, it can use the human language to communicate with machines or get feedback from machines in the form of a synthetic voice. Voice and language are the most ancient forms of communication between humans. Most people would prefer to interact with devices or machines using their voice and a spoken language instead of other methods such as typing or pressing buttons. For example, many people will prefer to listen to this book in what is known as an audiobook or prefer to start a voice call using voice commands on their cell phone.

Digital audio is therefore categorized into three main applications, as shown in Table 1.4.

TABLE 1.4
Digital Audio Application Types

Audio Type	Communication Direction	Description
Speech	Two-way	Voice interactive communications, voice commands, voice guidance
Media	One-way	Music playback, movie media, recorded speech
Hybrid	One-way and/or Two-way	Speech and media are used together or separately, which require one-way interaction and two-way interaction interchangeably

SPEECH AUDIO

When we refer to speech in digital audio, we usually refer to a form of communication that requires interaction. The simplest example of this form of speech is a telephone call. In a telephone call, one person is producing speech, while another person is consuming speech and may respond to it by producing speech back. This is a form of two-way interactive communication. Similarly, voice commands to a machine are also considered interactive, since the machine is required to respond to speech. In this case, the response may not be a return speech; instead, the response may be an action from the machine that the human expects to happen right away. An example of a voice command is a person asking the phone, "What is the weather right now?" and the phone responding back with a weather screen. The same is true for voice guidance systems. A simple example of a voice guidance system is navigation systems. In a navigation system based on GPS satellite communication, a machine provides guidance to a human on when to turn left or when to turn right in order to reach a destination. The person may be walking, riding a bike, or driving a car. The person may also climb stairs inside a building or climb a mountain on a hiking trail. In this type of situation, a machine is communicating using a synthesized voice, and the expectation is that the human would respond to the direction by making a movement in that direction, if this human wishes to reach the programmed destination. That means that an interaction is required. A two-way audio communication is therefore defined such as one way may be voice and the other way may be a non-voice reaction or a voice reaction.

MEDIA AUDIO

When we refer to media, we usually refer to recorded music or recorded voice and music. With media, we usually refer to one-way communication in which the human ear is passively consuming the recorded music or voice without a need to respond to the media source. The simplest example is listening to an Internet music streaming service. In this example, the Internet service is streaming recorded music content, which is sent the same way to millions of users. Different users may receive the same content at different times. Another example is Internet radio, where audio content is broadcasted to millions of people over the Internet. Although audio content may be produced at the same time, millions of listeners are not required to respond to it by communicating back to the Audio Source. It is worth mentioning that the Internet Broadcasted audio is prerecorded and buffered for a small portion of time, before arriving at the destination. This method is often referred to as store and forward. It means that different listeners of the Internet Broadcast may receive the content seconds apart from each other. No immediate interaction is required back to the broadcasting source, and, therefore, as long as the Broadcast is continuously sent in a timely manner, the listeners are able to consume the content, in what is known to be "live streaming."

In both live streaming and prerecorded streaming situations, a certain delay is allowed in the streamed media. The delay is compensated by buffering techniques in the media player of the client and server. In certain cases, the media is accompanied by other forms of digital content, such as video. When a person watches a digital movie, the person watches a moving picture that is synchronized with media audio. In this example, the expectation is that the person's ears consume the played audio media in unison with the eyes, which are consuming the moving pictures. In this case, there is a source-level interaction between the digital movie and digital audio; still, the person who consumes the audio content is not required to respond. In the case of audio–video synchronization, the delay is required to be smaller compared to the delay that is acceptable in pure media audio use cases.

HYBRID AUDIO

The third category of digital audio requires a mix of interactive voice and media. Consider, for example, an Internet game with multiple participants or a flight simulator. If we consider the digital audio that is generated in such a situation, then we are looking into media audio generated by the game server or flight simulator. In certain cases, the person who is consuming the audio is required to respond to the audio, by using either voice or other game or flight controls. This category is a hybrid of the first two in the sense that in certain periods of time, the content consists of one-way audio communication, and in other periods of time, a two-way communication is required.

One more example is an Internet gaming infrastructure in which gamers may speak to each other over an Internet connection while they are playing the game. In certain times, the gamers may passively listen to the game scene media sounds, and in other times, the gamers may begin an interactive conversation with other gamers. While the gamers are speaking, they may also hear back voice responses mixed with the game scene sounds. The result is hybrid one-way and two-way speech and media audio types.

THE SOCIAL ASPECT OF AUDIO

Digital Audio provides a social experience to people. We already reviewed the personal aspects of audio. But at its basics, audio is meant to provide interpersonal communication between different people. The communication may be between two people or between hundreds of people. For example, a lecturer is using speech audio, as verbal language, to convey messages to dozens or sometimes hundreds of people in a lecture hall. Another example is that people may gather together with family, friends, or coworkers to listen to news media on TV or to listen to music at parties. Social audio becomes very useful in public venues.

Places like train stations may use an announcement system to convey messages to the crowd. Places like museums may use audio as a digital guide to explain the exhibitions. And social places such as restaurants or bars may use TVs with news media or sports events for multiple people to listen and watch together. Many of these social audio interactions seem very natural to us. Many of these social audio interactions are happening today without the need for wireless audio or Bluetooth®. On the other hand, a very basic problem exists in public places – when there is a crowd of people in the same location. In these cases, not all people prefer to listen to the same audio content at all times.

This problem always exists, and the solution is sometimes left for people to agree on the volume level of the ambient audio and on what audio content to listen to. In some cases, people have control over the audio volume, while in others, the volume level is dictated by an administrator out of reach. Even when people have control over the audio volume in a public location, the problem is how to agree on the volume level selection among multiple people, each with a different preference. And then we also have to consider the audio content. Different groups of people may have different preferences on what audio to listen to. There are other cases where some people prefer to mute the audio content completely, which is against what other people may prefer.

The aspects of social audio, in terms of media content and volume levels, invite the need to use wireless audio as the Auracast™ Broadcast Audio Source. In social situations, certain groups of people may prefer to listen to content A at their volume level, and a certain number of people may prefer content B at their volume level. With wireless audio, these preferences may be met. For example, Figure 1.5 shows a picture of people in a restaurant with multiple quiet TV sets, while watching a game on a specific TV. A few TV sets in a sports bar may show different sports games or news channels. The TV sets are quiet, but use wireless audio to broadcast the media audio. People in the sports bar may tune their headsets to the TV of choice. This allows for social audio consumption of a group of people while still considering people who prefer to listen to something else or not listen to the TV at all, and have a conversation with a friend instead. Each person listening to a certain TV may tune the volume to the preferred volume for that person's headset. Different groups of people may tune to the same TV content while adjusting a different volume level in each headset. The person turning the volume down or up is not affecting other people's listening experience. A certain person may choose to answer a phone call and stop listening to the TV without affecting other people, who continue listening.

FIGURE 1.5   Social audio, people cheer in a restaurant, watching a game on TV.

The next few sections will provide a brief on Bluetooth in the context of audio. The next chapter (Chapter 2) will provide more details about the Bluetooth technology. This chapter provides a brief on Bluetooth from a historical and evolutionary perspective and how Audio evolved with Bluetooth.

AUDIO AND BLUETOOTH® TECHNOLOGY

Radio communication was known since the 1890s when radio waves were discovered. Radio waves carry analog audio across medium to large distances. Initially, the radio frequency served as a carrier frequency on which the analog audio acoustic wave was modulated using either amplitude modulation or frequency modulation.

Wireless is a term that is related to the revolution of carrying digital data or digital audio over radio frequency. Digital wireless communication theory emerged in the mid-20th century, but developed commercially around the 1990s. The commercial digital wireless revolution began with cellular communication systems moving from analog communication to digital communication – soon followed by short-range technologies such as Wi-Fi and Bluetooth®. With digital wireless, both voice and data are considered as buffers, as bits, which represent a conversion of digital forms of media. The radio waves are carrying or modulating bits instead of an analog wave. When the wireless revolution began with data and digital audio communication, there became a need to communicate over short distances. There was a realization that wireless communication across the room is essential for peripheral devices. Wi-Fi, for example, enabled a portable laptop to connect to a local network in the room or a nearby hub. The first need for Bluetooth wireless technology was to connect a personal computer to a phone. The other initial Bluetooth application was making a phone call without the need to connect a long trail of wire from an earbud to the phone.

Bluetooth wireless technology provides a personal experience to people. Over the years, Bluetooth has evolved into a companion in our day-to-day activities. We use Bluetooth with our peripherals and gadgets, in our personal computers, on our phones, and on the move while driving, biking, running, or walking. Bluetooth allows us to use technology without the need for wires. Wireless means that the person using the technology has more freedom and convenience. The person may walk around while wearing or holding Bluetooth-enabled equipment.

The fusion between Bluetooth technology and audio provides the ultimate personal experience for people. It is also the fusion between acoustic signal processing and radio frequency signal processing. The acoustic signal part is the content that the person uses to create and enjoy audio, and the radio frequency signals are the medium on which the audio content travels from one point to another. Bluetooth allows you to enjoy music or have phone calls without the need to connect a wire between headsets and an Audio Source.

From its inception, in the mid-1990s, Bluetooth technology was targeting audio as one of its main use cases – initially, voice calls and later followed by music streaming. The fusion between audio and Bluetooth is one of the main reasons for the popularity of Bluetooth.

CLASSIC AUDIO EVOLUTION

Initially, Classic Audio evolved to fill gaps. When Bluetooth® technology was first introduced in the mid-1990s, voice calls over cellphones were the killer applications. So the main audio task for Bluetooth was to provide the ability to connect an earbud to a cellphone without a wire. The first version of Bluetooth only addressed the voice call use case. And the audio quality was modest. The Codec was compressing voice calls using a sampling rate of 8000 Hz, with a waveform Codec. In the 1990s, this was considered good enough. However, it was known that a sampling rate of 8000 Hz does not cover the entire human voice spectrum. An 8000 Hz sampling rate covers a spectrum content of 4000 Hz. 4000 Hz includes most of the human voice tones, but higher tones are left out, especially certain tones, such as "shhhh," or higher tones, including the voices of women or children.

The Classic Audio technology to send and receive voice did not take into account all aspects of error detection, such as retry. This gap caused click and pop sounds during a voice call whenever radio interference caused an error in the received speech frame. In later versions of the Core specifications, Classic Audio filled the gap with error detection and retries, and the support for a wideband speech Codec, with a sampling rate of 16,000 Hz.

Accordingly, the Host profiles initially only defined signaling for call control. The protocols to handle call control were adopted by Bluetooth from other non-Bluetooth specifications to save development time and ease the market adoption of Bluetooth by using an existing known technology at the time.

WHAT ABOUT MUSIC?

Initially, use cases and profiles were developed for voice only. Later, Bluetooth® technology filled a new gap that emerged in the 2000s, which is music playback on the go. Listening to music with a wired headset was popular since the 1970s with portable cassette players. Later, when music turned digital, it was possible to listen to music from PCs and phones, using speakers or wired headphones. As Bluetooth provided a solution for voice audio, a new gap emerged to also support wireless headphones' music playback. However, the original architecture of Bluetooth® Classic was targeting voice only. The codecs and transports supported bitrates that could only cover the voice bandwidth. And the protocols and profiles were only addressing call control.

The solution back then was to define a new architecture to address music only. As a result, two separate transports were used for voice and music. A music stream in Classic Audio uses a different transport than a voice stream.

Also, new profiles were developed for music and media playback, which may only use the music transport. New codecs were defined to support music sampling rates of 48,000 Hz or 44,100 Hz, and separate transports were configured to support the required bitrate.

CLASSIC AUDIO APPLICATIONS

The control and signaling messages for music and voice use a different set of protocols in Classic Audio. Handling volume is done separately for voice and music in Bluetooth® Classic audio. Over the years, products have allowed the usage of both voice and music. No more earbuds only for voice calls and headphones only for music playback. Wireless headsets emerged with built-in microphones, which allowed using the same device to answer a call or to begin playing a music track. Car kit systems allowed answering calls while driving without holding the cellphone, and these car kits also supported music playback from a playlist on the phone.

The voice and music in Bluetooth Classic use a different architecture, so the vendors had to find creative ways to support both music and voice – oftentimes with additional nonstandard controls to make sure the voice stream is stopped before a new stream is established from the music profile. This is often called the break-before-make approach, where one use case stops before another begins. Some other vendors chose the opposite direction, to make-before-break, which means, for example, to set up music and then stop voice. Since Bluetooth Classic uses two different architectures for voice and music, such a nonstandard approach needed a lot of trial and error and was successful sometimes and failed in other times.

The main concern is that end-user interoperability was not well defined. As a result, the end-user experience was compromised when products from different vendors interacted. It works most of the time, but there is no assurance since no standard exists on how to switch between calls and music.

Another nonstandard use case that was developed to fill a gap was the need to send stereo to different devices on two ears. This is also known as true wireless. In true wireless, instead of a headset, two different Bluetooth® devices are receiving audio from one source, which may be a PC or a phone. Also, in this use case, vendors had to find creative ways to solve this problem. This problem was left open by the Classic Audio standard, because it only defined a protocol for point-to-point audio streaming. The solution was to use Bluetooth to stream music to one earbud and let the other earbud sniff the air and capture the stream. This way, each earbud could play either the left content or the right content of the stereo audio. The solution required out-of-band signals between the earbuds and a remote control unit that often also served as the case for charging the earbuds.

CLASSIC AUDIO LIMITS

Classic Audio limits were broken when nonstandard solutions appeared. However, the nonstandard solutions still needed Bluetooth® technology as the skeleton for interoperability. There are more use cases to support sending audio to more than two receivers, such as surround sound systems, or to broadcast audio to a big crowd of people. These new use cases are beyond the limits of Classic Audio, even with nonstandard workarounds.

As we reviewed with you, Classic Audio was developed to fill gaps. Classic Audio was successful in providing solutions to basic voice and media applications. To a certain extent, this was a good enough approach. Looking forward into the future of digital wireless audio, it became clear that Classic Audio reached a barrier. Applications for hybrid audio are limited in what capabilities they enable with Classic Audio. And social audio applications are nearly impossible with Classic Audio.

THE ONE ARCHITECTURE

LE Audio development by the Bluetooth® SIG is addressing use cases in a single unified architecture. The objective of LE Audio is to define the One Architecture as a standard that enables interoperability and promotes a better user experience.

When we refer to the One Architecture, a number of basic audio utilities join together to provide full coverage of audio use cases. One Architecture for the stream and audio transport for either voice or media. One Architecture to handle audio to and from multiple devices and to publish and share audio, via Unicast or via Auracast™ broadcast audio. One Architecture to define Codec settings and quality of service. One Architecture to configure audio, control volume, answer calls, or play a media track.

The advantage of One Architecture is a better user experience, stretching the limits beyond what's possible with Classic Audio, and also looking forward to enabling new use cases with greater ease.

The road for the One Architecture is paved using LE Audio, and a new audio experience awaits along every stop.

In the next chapter, we will review the details of Bluetooth technology before delving deeper into the LE Audio One Architecture.

2 Bluetooth® Overview

In this chapter, we will provide a brief overview of Bluetooth technology, as it evolved from Bluetooth® Classic to Bluetooth® LE. We will also review some of the concepts in Bluetooth Classic that are used by Classic Audio and the Bluetooth LE concepts that are used by LE Audio. The focus in this chapter will be more on the Bluetooth radio technology rather than audio. The next chapter will return the focus back on LE Audio.

BLUETOOTH®

Bluetooth is a short-range wireless technology. The range of Bluetooth is typically less than 10 meters. Bluetooth is also known as IEEE standard 802.15.1 but is no longer maintained by the IEEE organization. Bluetooth is managed by the Bluetooth SIG, which has more than 35,000 member companies from various different fields – like semiconductors, mobile phones, personal computing, consumer electronics, automotive, and so on. A manufacturer of a Bluetooth device must meet Bluetooth SIG standards to market it as Bluetooth-compliant equipment. Bluetooth uses short radio wavelengths in the unlicensed spectrum, known as the ISM band, which is reserved for Industrial, Scientific, and Medical purposes. The radio band in use by Bluetooth ranges from 2400 to 2483.5 MHz.

Bluetooth® Classic refers to the main evolution of Bluetooth technology between 1998 and 2010. In 2010, an additional technology was added to Bluetooth – Bluetooth® LE. Bluetooth Classic and Bluetooth LE continued to evolve as separate radio technologies that share many concepts and differ in other concepts. Bluetooth LE is reviewed later in this chapter.

RADIO MODEL

Bluetooth® technology uses radio technology called Frequency Hopping Spread Spectrum (FHSS). The available band from 2400 to 2483.5 MHz is divided into 79 subbands, starting from 2402 MHz with a width of 1 MHz each. The first RF channel is 2402 MHz, and the last RF channel is 2480 MHz. Bluetooth technology is a packet-based protocol and transmits each packet in one of the 79 designated Bluetooth channels, which are also called hops. The Bluetooth devices rapidly hop, 1600 times per second, over these channels in a pseudorandom pattern that is algorithmically determined by certain fields of the address of the Bluetooth device and the network clock of a Central device. This mechanism is used to avoid interference and also to prevent eavesdropping to a certain extent. In fact, FHSS was invented in WWII as a military audio encryption technique. Bluetooth technology, however, does not rely on FHSS for encryption and employs an additional encryption over the data using dedicated encryption private keys, which are authenticated and only known to the network devices. Every hop out of the 1600 hops per second may be an opportunity for communication. In practice, a small subset of the 1600 hops is used by a single device for transmission or reception.

DOI: 10.1201/9781003590187-2   **19**

MEDIA ACCESS CONTROL

Bluetooth® technology uses a Central-Peripheral architecture. One Central may communicate up to seven Peripherals in a piconet. A Bluetooth device may participate concurrently in two or more Piconets, but it can never be a Central of more than one piconet. In a given piconet, the Central always transmits first, addressing a certain Peripheral, and only this Peripheral is allowed to respond. A Peripheral cannot send data unless the Piconet Central device addresses it. The Piconet Central takes turns to communicate with all Peripherals in the Piconet.

The Central provides a base clock for the piconet for packet exchange; the clock is a counter of a unit half slot time. The Central clock ticks with a rate of 312.5 µs (reminder, µs stands for microseconds). One slot is made up of two clock ticks, that is, 625 µs. The slot duration of 625 µs determines the hop frequency, since there are 1600 slots of 625 µs in 1 second. The Central clock ticks also when slots are not used for sending or receiving data. Packets may be one, three, or five slots long, and in all cases, the Central transmits at even slots and the Peripheral transmits at odd slots. Once a packet length is selected, the first hop slot tick determines the 1 MHz frequency band to use for the entire packet. The entire packet is therefore sent or received over a single RF channel. RF channels never change within a packet. The next packet selects a different hop frequency based on the clock tick and a 48-bit MAC address of the Central. The Central and Peripheral share the information about the Central MAC address and its clock, when creating the connection, so the next hop frequency is always known by all Piconet participants. The MAC address in Bluetooth is also known as BD_ADDR (Bluetooth Device Address).

Figure 2.1 shows an example of communication between the Central and Peripheral using single slots, three slots, and five slots. The slots are shown as dotted lines. Each slot is labeled with a slot number (e.g., k, k+1, k+2); the incrementing k means the next pseudorandom RF frequency for this slot number. When a packet begins with a certain slot number, then the RF channel calculated by this slot tick is used to transmit the entire packet, and also when it spans over multiple slots. The length and duration of the packet are stated in the packet header.

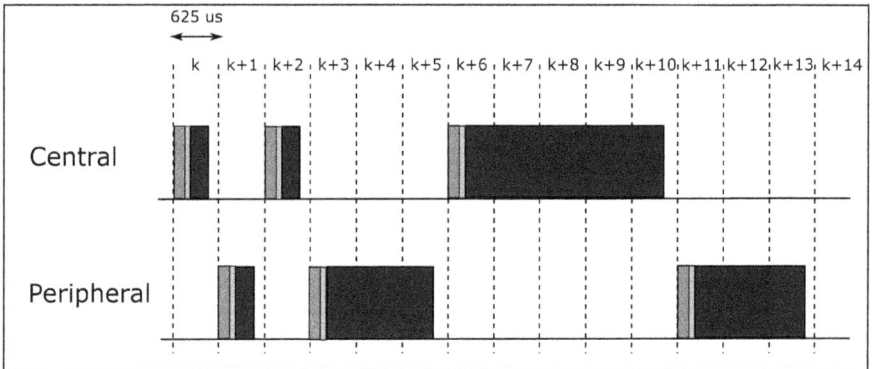

FIGURE 2.1 Central and Peripheral packet transmission architecture.

Physical Layer Modulation Types

The process of imposing an input onto a carrier wave is called modulation. In the case of Bluetooth® technology, the input to modulate is the data bits. Bluetooth uses two types of modulation – GFSK (Gaussian Frequency-Shift Keying) and DPSK (Differential Phase-Shift Keying). Based on these modulation techniques, two types of rates are defined. The maximum physical symbol bitrate for **Basic Rate** (BR) modulation (which uses GFSK) is 1 Mbps. The maximum physical symbol bitrate for **Enhanced Data Rate** (EDR) modulation (which uses DPSK) is 3 Mbps. EDR also supports 2 Mbps PHY. Overall, there are three PHY rates in Bluetooth: 1 Mbps (GFSK), 2 Mbps (2-EDR, 4-DPSK), and 3 Mbps (3-EDR, 8-DPSK). GFSK is a binary modulation in which the radio modulates a single bit by either a positive or a negative frequency offset. 4-DPSK modulates four different phases and therefore modulates 2 bits per symbol (2 to the power of 2). 8-DPSK enables eight phases and therefore modulates 3 bits per symbol (2 to the power of 3). The GFSK modulation is the most robust, but produces a lower bitrate. The GFSK modulation is a constant envelope, which means that the symbol is modulating frequency using a Gaussian filter, which prevents sharp changes to frequency. On the other hand, 4-DPSK and 8-DPSK are non-constant envelope modulations. The DPSK modulation family involves sudden phase changes from one symbol to the next, which makes amplitude changes to the radio signal envelope. Constant envelope modulation, such as GFSK, has an advantage in terms of low power consumption as the amplitude level remains constant, and the peak-to-average power ratio (PAPR) remains constant from one symbol to the next, due to smoother frequency offset changes. A constant zero changing PAPR consumes less battery power, since it may use a nonlinear power amplifier, due to the fact that it operates around a given constant amplitude. Otherwise, for non-constant envelopes, a linear power amplifier is required, which consumes more power, involves a higher cost, and requires more calibrations to remain linear. The above means that there are use cases and devices that would prefer to use GFSK and lower bitrates, in order to save more power, and keep the cost of the radio component lower.

The practical maximum MAC bitrates are smaller than the physical data rates. This is because MAC packets contain header parts for addressing and trailer parts for error detection, such as CRC (Cyclic Redundancy Check), as previously shown in Figure 2.1 at a high level, and later shown in Figure 2.4 in more detail. The packets also always occupy part of a slot to allow transition of the RF transceiver from TX to RX and back, as also shown in Figure 2.1. There is always at least around 200 μs of spacing left from the end of one packet on channel k to the beginning of the next slot on channel k+1.

Due to the MAC overhead of packet fields and gaps between packets, the maximum practical MAC BR rate is 0.72 Mbps, and the maximum practical MAC EDR rate is 2.2 Mbps for 3-EDR and 1.45 Mbps for 2-EDR. The MAC scheduler may not use all transmit and receive opportunities, depending on other activities, so the practical application throughout will be lower, depending on the type of device and the type of Bluetooth application in use. As we discussed before, different applications and devices may use a subset of the available spectrum in a reduced duty cycle. As is common to many communication technologies, the applications themselves may use protocols that add additional header/trailer overheads on top of the actual application data.

LOGICAL LINK CONTROL

For communication, a physical radio channel is shared by a group of devices synchronized to a common clock and frequency hopping pattern. There is a hierarchical layering of links and channels above the physical channel, which is used to create logical communication between any two Bluetooth® devices. The hierarchy is physical channel, physical link, logical transport, logical link, and L2CAP channel (Logical Link Control and Adaptation Protocol). In order to support multiple concurrent operations, the devices use time division multiplexing between the channels.

There are special designated physical channels in Bluetooth – Basic piconet, Adapted piconet, Inquiry scan, Page scan, and Synchronized scan channels.

The Basic piconet channel is used for communication between connected devices during normal operation.

The Adapted piconet channel is the same as the basic piconet channel, but the primary difference is that it does not use all of the 79 frequencies and also the Peripheral transmits in the same frequency as the Central in the previous transmission – that is, frequency is not recomputed between Central and subsequent Peripheral packets. The purpose of the Adapted piconet channel is to avoid hopping in channels that are classified as noisy channels; the protocol to determine and exclude noisy channels is known as AFH (adaptive frequency hopping). When AFH is activated, the Peripheral always responds with the same RF channel that was selected by the Central.

The Inquiry scan channel is used by a device to be discoverable to other devices, and it responds to inquiry requests from other devices on this channel.

The Page scan channel is used by devices to be connectable to other devices and responds to page requests from other devices on this channel.

The Synchronized scan channel is used to recover the current piconet clock as the synchronization train packet is scanned by the scanning device for this purpose.

Figure 2.2 illustrates the layering of the various channels and links.

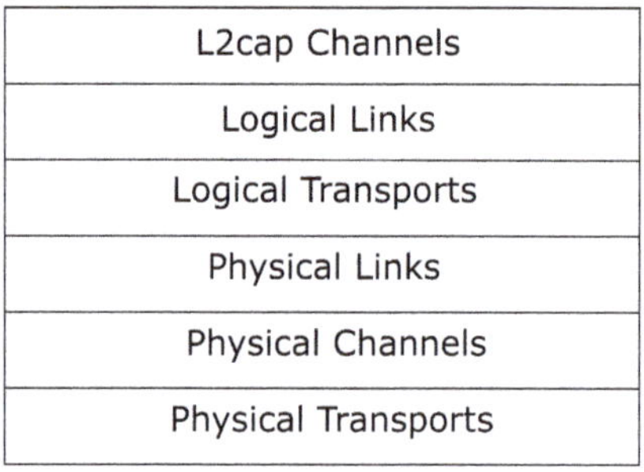

L2cap Channels
Logical Links
Logical Transports
Physical Links
Physical Channels
Physical Transports

FIGURE 2.2 Overview of transport architecture.

THE MAC SCHEDULER

The decision of what hops to use and when to use the hops is a decision made by what is known as the Bluetooth® scheduler. The Bluetooth scheduler follows the MAC protocols, which in Bluetooth is also known as the Baseband and the Link Controller. The MAC policies and rules in the Baseband and Link Controller are mostly derived from a set of protocols defined by the Bluetooth Core Specification. The set of protocols defines how and when it is allowed to access the air media for transmitting or receiving. To a certain extent, the Bluetooth scheduler is also allowed to take autonomous decisions about which air activity to participate in. Bluetooth specifications define that multiple activities may be deployed simultaneously by a given device. A device may choose not to transmit or receive at all, in order to save power, or it may choose to use a subset of the 1600 hops per second to participate in communication with another device. The device may also use some of the 1600 hops per second for discovery purposes, or make itself discoverable. The decision as to what extent the 1600 hops per second are used is left to the Bluetooth Classic scheduler in any given device, as long as it conforms to the protocols in the Bluetooth Core Specification.

The decision of what activities to enable is driven by a Generic Access Profile (GAP), which is in turn driven by other High-Level Profiles and applications in order to achieve end-to-end connectivity. The Bluetooth Core Specification is defined to enable a level of flexibility on how the 1600 hops per second are used and also allow a device to use a small subset of these hops in order to save more power when communication is not required or smaller amounts of data are needed. As a result, when you are analyzing the air traffic used by a specific Bluetooth device, you will discover that most of the air is unused, and from time to time, you will observe bursts of traffic. Some devices are generating longer bursts of traffic, and some devices are generating very short bursts of traffic, so the effective duty cycle of the spectrum may range from around 10% to around 40%. This is known as application throughput, and it allows multiple close-by Bluetooth devices to share the radio spectrum by selecting different channels and different times for transmission. The diversity is in time and in the frequency space, which enables multiple devices to transmit with a smaller collision probability.

Figure 2.3 shows an example of a short time slice with Bluetooth Classic traffic bursts. This example represents a zoom-in on a 12 ms duration. Two Piconets are shown, and in each Piconet, a communication between one Central and one Peripheral. An example of such a topology could be two people each using a cellphone connected to a headset over a Bluetooth connection. The two people may be listening to music over Bluetooth. The figure shows three axes as three-dimensional planes: distance axis, time axis, and frequency axis. Each Piconet chooses a frequency via a pseudorandom sequence as derived by the Piconet Central devices. Figure 2.3 shows that frequencies are selected statistically such that there is little overlap between the two Piconets. In each Piconet, the Central is sending first, and the Peripheral responds. The traffic may be a single slot, three slots, or five slots. The figure shows two cases. In the top case, AFH is disabled; in this case, the Central and Peripheral send packets on different RF frequencies. In the lower case, AFH is enabled, and certain RF frequencies are excluded, while the Central and Peripheral are using the same RF frequencies in each exchange.

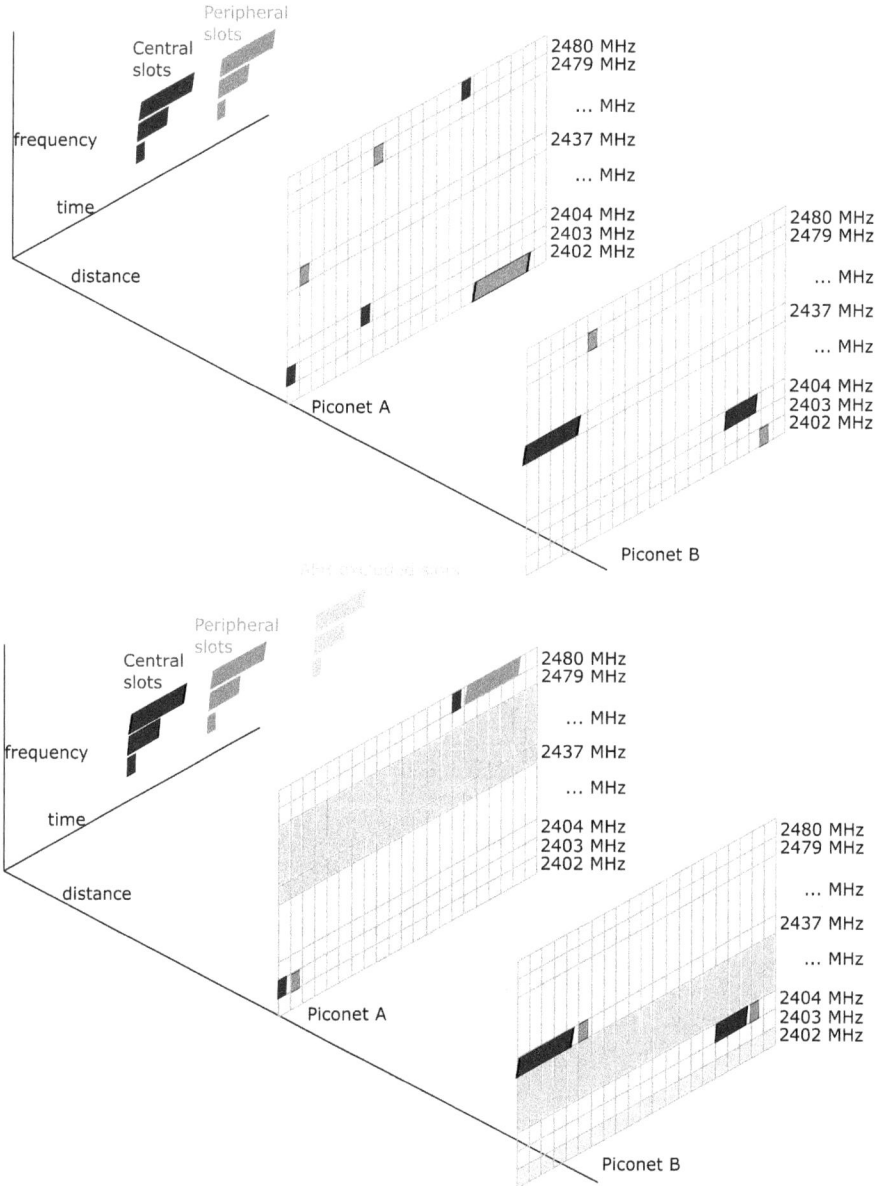

FIGURE 2.3   Spectrum usage examples for non-AFH and when AFH is enabled.

In the top part of Figure 2.3, there is an example for a possible collision in frequency 2437 MHz, in which the Central in Piconet B sends a five-slot packet at the same time when the Peripheral in Piconet A sends a single-slot packet on the same channel. This collision may cause the Central in Piconet A to miss the packet, depending on the distance from Piconet B and the power levels of each device. If this happens, the Central in Piconet A will request the Peripheral to retry the packet, as shown by another packet in channel 2403 MHz from the Central and a response in channel 2479 MHz from the Peripheral. Piconet B may be using a higher power level and was not impacted by the single-slot packet from Piconet A on channel 2437 MHz, and no retry was done. Note that Piconets A and B are not time aligned, as each Piconet is aligned to a clock from a different Central device. Also note that, depending on the distance and the power levels, overlapping channels do not always result in collisions and lost packets between Piconets.

In Figure 2.3, the lower part shows that when AFH is enabled, each Piconet excludes certain RF frequencies. This is also known as the AFH channel map. The Central may update the channel map based on the channel quality or knowledge about the usage of RF frequencies by other technologies such as Wi-Fi. The Peripheral may classify channels and report on the RF channel quality for each RF frequency. Based on the information on RF channels, the Central may adaptively update the AFH channel map to avoid interference. The lower part shows that each Piconet adapted to a different AFH channel map based on local interference close by to that Piconet – which aids in collision avoidance – since Piconet A excluded frequencies 2437–2479, which are free for Piconet B to communicate, and similarly Piconet B excluded frequencies between 2404 and 2437, which are free for Piconet A to communicate. In the lower part, it is shown that when AFH is enabled, RF channel 2437 MHz is not used by Piconet A, and the previous collision example in the upper part is therefore avoided, while the Peripheral is following the same RF channels used by the Central (and the Central did not initiate a retry request).

PACKET STRUCTURE

Figure 2.4 describes the packet format for both the BR and the EDR modulation. The packet is divided into three main parts.

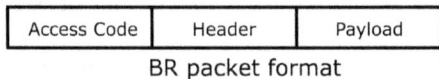

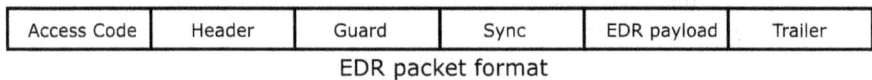

Access Code	Header	Payload

BR packet format

Access Code	Header	Guard	Sync	EDR payload	Trailer

EDR packet format

FIGURE 2.4   Packet format for BR and EDR modulation.

The first part in the packet is the Access Code. The Access Code begins with a short 4-bit preamble of alternating ones and zeroes, followed by 68 bits that form a unique code. The code is used to determine the media access control rules. In discovery phases, the Access Code is derived from a fixed global address or from the 48-bit address of the device that makes itself connectable. In a Piconet, the Access Code is derived from the 48-bit device address of the Central. All Peripheral devices correlate to the Access Code to determine if they received communication from the Central device. The Access Code determines the physical channel.

The second part of the packet is the Header that contains addressing fields and information about the payload, such as packet type and logical link control, and sequencing retry logic. Packet types may be various ACL types (Asynchronous Connection) and SCO/eSCO types (Synchronous Connection Oriented), or other types such as POLL or NULL. The Header determines the physical link between a Central and a Peripheral. In a Piconet, the Header determines which Peripheral may transmit back to the Central. The Header is protected by a dedicated 8-bit CRC checksum field (Cyclic Redundancy Check), which is called HEC (Header Error Check) and coded via FEC 1/3 (Forward Error Correction).

The third part of the packet is the Payload. In EDR, there is a guard and sync period before the payload begins, and a trailer when the payload ends, because the Access Code and Header are always sent in BR GFSK modulation. The guard and sync allow the transmitter and receiver to switch from BR to EDR, between the Header part and the Payload part, and sync at the end of the modulated payload. The Payload contains a 16-bit CRC field, so the receiver may calculate the checksum and determine if the payload contains errors. The payload is encrypted in most use cases, which mandate privacy. Certain packet types may also use FEC on the payload for better robustness in certain link quality conditions.

Low-Power Modes

There are low-power modes defined in BR/EDR that are used to specify a low-duty-cycle operation consisting of modified periods of absence and presence of devices. The low-power modes are Hold and Sniff modes.

The Hold mode allows scheduling a one-time absence period from the link. The Hold mode is hardly used by applications since it has a negative effect on user experience.

The Sniff mode allows the reduction of the duty cycle of ACL connections and the negotiation of periodic listening intervals, where devices may communicate. The Sniff mode is widely used by applications when the device moves to a standby operation and may save more power by waking up for a shorter period of time. When applications get out of standby, they typically request to switch from Sniff mode to active mode, so control signaling and application data can move faster. Certain applications use the Sniff mode as a mechanism to secure guaranteed quality of service if they need low latency and low bitrate. Examples for such applications are HID Bluetooth® devices like mice and keyboards (Human Interface Device). HID devices send a low bitrate of cursor movement or keystrokes, but they need a low-latency response for human eye interaction. Therefore, HID devices negotiate a

Sniff interval that matches the low-latency response and a window of communication which matches the low bitrate. An example for a HID keyboard sniff configuration is an interval of 15 ms, which are 24 slots, and a window of up to 6 slots to communicate the character keystrokes.

HOST AND CONTROLLER

A Bluetooth® protocol stack is divided into two main parts – Host and Controller – and there is a standardized layer of communication between the Host and the Controller called the Host Controller Interface or HCI (Figure 2.5). The Controller part of the stack typically resides in the Bluetooth chipset, and the Host resides in the Application Processor, while HCI is carried over a bus transport between the Host and the Controller. The standard HCI allows the implementation of the Host by one vendor and the Controller by another vendor.

The standard HCI allows manufacturers to use the same Bluetooth chipset that implements the Controller part and the Host Application Processor. This is typically the case in small Bluetooth devices such as headsets, earbuds, and mice. In this case, HCI becomes a functional interface as opposed to a physical bus, and the Host and Controller share the same chipset.

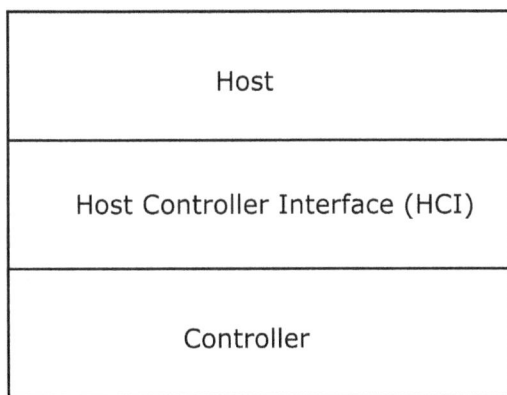

Host
Host Controller Interface (HCI)
Controller

FIGURE 2.5 Host Controller Interface.

The Bluetooth host begins with L2CAP – Logical Link Control and Adaptation Protocol. L2CAP serves a few purposes. It provides the basic framing of application data and control over a point-to-point connection between two devices. The L2CAP protocol allows opening different channels to serve different application profiles for control or to send data. The L2CAP protocol allows multiplexing a single connection between multiple applications by providing a channel identifier to each application. The L2CAP protocol allows configuring a data channel to provide a certain quality of service for low latency, isochronous, or best effort.

Above L2CAP, a set of protocols and profiles enables a variety of use cases.

We will refer to Bluetooth as Bluetooth® Classic as we delve into the next generation of Bluetooth called Bluetooth® LE.

BLUETOOTH® LOW ENERGY

Bluetooth Low Energy (BLE or Bluetooth® LE or Bluetooth Smart) came into existence from Bluetooth version 4.0 onward.

SIMILARITIES AND DIFFERENCES TO BLUETOOTH® CLASSIC

Bluetooth® LE operates in the same 2.4 GHz ISM band as Bluetooth® Classic, and the initial target applications were primarily those that do not exchange large amounts of data and can therefore run on battery power for a long time – even years. As the name suggests, Bluetooth LE is designed for low battery consumption, and it is achieved by keeping the Bluetooth LE device in sleep mode most of the time.

The protocol stack is also divided into Host and Controller with a similar Central-Peripheral architecture; except that there is no limit to the number of Peripherals connected to the Central. Unlike BR/EDR Peripherals, Bluetooth LE Peripherals do not share a common physical channel with the Central. Instead, each Peripheral communicates on a separate and independent physical channel with the Central. Bluetooth LE introduces the data Advertising concept, which does not exist in Bluetooth Classic. While Advertising is used to discover and connect devices, it also enables connectionless operation in which a single device can advertise general application data to an unlimited number of listeners. Advertising enables data distribution by investing a small amount of power, allowing multiple devices to scan and receive the data. The advertisers may be small sensors running on coin cell batteries.

Bluetooth LE is not compatible with Bluetooth Classic, but a Dual Mode chipset may be designed with both Bluetooth LE and Bluetooth Classic technologies. Bluetooth LE also uses the same FHSS mechanism as Bluetooth Classic. The basic RF channel hopping in Bluetooth LE is slower compared to Bluetooth Classic. In Bluetooth LE, it is possible to stay on the same channel for a few exchanges of transmit and receive between two devices. Therefore, regulatory bodies often treat Bluetooth LE as a Direct Transmission System (DTS) or Hybrid System (where both FHSS and DTS are used). The other reason for treating Bluetooth LE as non-frequency hopping is that only three channels are used for discovery, and that the number of channels in the first Core version of LE was allowed to drops down to 2 (where at least 15 channels are typically considered FHSS). Later versions of the Bluetooth Core Specification added a faster channel hopping mode and usage of a wider set of discovery channels.

RADIO MODEL

Bluetooth® LE uses 40 channels with a channel spacing of 2 MHz. For comparison, with Bluetooth® Classic, this is about half of the 79 channels with channel spacing of 1 MHz each. Although the channel spacing in Bluetooth LE is 2 MHz, the actual bandwidth in use by default is 1 MHz, which leaves more spacing between adjacent channels. This choice was made to simplify the radio design of tiny Bluetooth LE devices. Bluetooth LE allows an optional 2 MHz PHY as well to enable higher bitrates.

PHYSICAL LAYER MODULATION TYPES

Bluetooth® LE uses GFSK modulation. The maximum symbol physical bitrate for Bluetooth LE is 1 Mbps when a 1 MHz channel width is used and 2 Mbps when using a 2 MHz channel width, at the expense of range. Alternatively, it is possible to use coded PHY in Bluetooth LE and achieve up to four times the range at the expense of data rate. Coding reduces the symbol rate to 500 kbps or 125 kbps. The practical maximum MAC bitrate is 1.4 Mbps. The actual application rates in Bluetooth LE are lower, since the radio is off most of the time for most of the application types.

LOGICAL LINK CONTROL

Just like BR/EDR, there are designated Bluetooth® LE physical channels that are defined, and they are Piconet, Advertising, Periodic, and Isochronous Channels.

The Bluetooth LE Piconet physical channel is used for communication between connected devices and is associated with a specific piconet. There are 37 Bluetooth LE Piconet channels.

The Bluetooth LE Advertising physical channel is used for broadcasting advertisements to Bluetooth LE devices. These advertisements can be used to discover, connect, or send user data to scanner or initiator devices. There are three primary Advertising channels and 37 secondary Advertising channels used for extended advertisements.

The three primary Advertising channels are used for the transmission and reception of Advertising packets of various types, and to carry Broadcast data for applications, and to discover and connect to other Bluetooth LE devices.

The remaining 37 channels are Bluetooth LE data channels and are used by devices to send data to each other after the connection is established. Bluetooth LE data channels are used by either Central or Peripheral devices for the transmission and reception of data packets. The remaining 37 channels are also used as secondary Advertising channels for sending connectionless Advertising data from the Host or from the Controller. The secondary Advertising channel is used to offload data that would otherwise be transmitted in the primary Advertising channel. Sending Advertising data in secondary Advertising channels is considered more efficient as more channels are available, different PHY rates are possible, and less repetition of data is done.

The Periodic physical channel is used to send user data to multiple devices in Periodic Advertisements at a specified known interval, without the need for remote devices to scan repeatedly, and therefore conserve power.

The Bluetooth LE Isochronous physical channel is used to transfer Isochronous data between Bluetooth LE devices in the Bluetooth LE connection or to transfer Isochronous data between unconnected Bluetooth LE devices via a connectionless Broadcast channel. The Bluetooth LE Isochronous physical channel was introduced in Bluetooth Core 5.2 for the use of LE Audio and will be a special emphasis of this book in later chapters.

THE MAC SCHEDULER

As in BR/EDR, the MAC scheduler is configured by GAP to either scan, advertise, or create connections with a given interval and duty cycle. A given device may support multiple activities, and the MAC scheduler executes the activities on a time-sharing basis while reserving bandwidth for scan, Advertising, or connection activities. The MAC scheduler takes turns to scan, advertise, or participate in a connection. Unlink in BR/EDR, the communication between devices is not scheduled in time slots. Instead, a default inter frame space of 150 µs exist between consecutive packets (this value may be negotiated to a lower duration between Central and Peripheral). This results in a more efficient use of the spectrum because packets are not constraint to begin in fixed 625 µs slots like in BR/EDR.

In Bluetooth® LE, most activities are scheduled based on deterministic rate scheduling, which means that ACL connection, for example, is configured via an interval and a window (the window is called a connection event). In ACL, a connection has a connection event interval and a connection event window. Similarly, Advertising is configured with an Advertising event interval and an Advertising event window. The advantage of this approach is that bandwidth for different activities is reserved and the quality of service is guaranteed while keeping power consumption low, since devices know exactly when to wake up in order to receive data in a connection. The same concept is also extended to Periodic Advertising and to Isochronous Channels (CIS and BIS). The exception to this rule is when a device is scanning for Advertising. When scanning for Advertising, the scanner is initially unaware of what the interval used by the advertiser is. Therefore, the scanner invests more energy to keep the receiver on until it is able to detect Advertising. This is the reason why Bluetooth LE profiles define the Central devices to be the scanner and the peripheral devices to advertise. There are cases where peripheral devices are required to synchronize to remote advertisers, such as Periodic Advertising, which contain timing and configuration of a Broadcast transport. In these cases, there are dedicated link layer procedures to transfer the synchronization information from centrals to peripherals, so peripherals may skip scanning and synchronize to the fixed Periodic Advertising interval immediately.

The Bluetooth LE MAC scheduler is therefore considered more power-efficient compared to a classic MAC scheduler, since it is configured as low power mode from the get-go instead of having to switch to low power, such as in the classic BR/EDR MAC scheduler.

PACKET STRUCTURE

Figure 2.6 describes the Bluetooth® LE packet structure. The Bluetooth LE packet is divided into four main parts. The first part of the packet is a preamble, which is a fixed pattern of alternating ones and zeroes. The length of the preamble in bits is different for the different Bluetooth LE PHY types: 1M, 2M, Coded PHY, and 2M 2BT. LE 2M 2BT PHY is a new physical layer configuration introduced in Bluetooth® Core 6.0, specifically designed for use with the Bluetooth Channel

Sounding feature. This PHY type enhances the capabilities of Bluetooth LE by providing a high symbol rate and improved performance in distance measurement applications. The purpose of the preamble is to allow the receiver to tune the received RF gain and adjust the timing offsets and frequency offsets.

The second part of the packet is the Access Address. The Access Address is a unique number that is either fixed for Advertising (primary or secondary), and discovery, or randomly assigned by a Central or a Broadcaster in the case of a specific physical link. The various Access Addresses for specific physical links include PA (Periodic Advertising), ACL (Asynchronous Connection), CIS (Connected Isochronous Stream), or BIS (Broadcast Isochronous Stream). The Access Address, therefore, describes both physical transport and physical link. After the Access Address is received by the target device, the receiver knows how to parse the Protocol Data Unit (PDU).

The third part of the packet is the PDU, which may carry either Advertising data or physical link data, depending on the Access Address. The PDU contains a PDU header, which has a different format depending on the physical link, as determined by the Access Code: Advertising, PA, ACL, or Isochronous PDUs (CIS header or BIS header). The Data channel PDU may be either ACL Control, ACL data, or Isochronous data. Data PDUs are encrypted by the Controller, and an MIC (Message Integrity Check) is added as a trailer to the PDU. Advertising PDUs (PA or general Advertising) are not encrypted and contain nonsensitive Host Advertising data or hashed Host data, which is encrypted via higher Host protocols.

The fourth part of the packet is the CRC, which calculates the checksum on the PDU parts (including the PDU header). The result of this checksum helps the receiver determine if the packet received is without errors. In the case of Advertising, packets with errors are dropped. The Advertising procedure contains redundant packets, so a subsequent packet may be received correctly. In the case of Data PDUs, packets with errors will trigger the receiver to send a request for a retry of the packets, as part of the return PDU packet header. A special case of Data PDU is BIS PDU, which behaves similarly to Advertising since it is a connectionless physical link with no feedback path from the receiver. BIS PDUs are sent in multiple copies, so in case one PDU contains a CRC error, the next PDU may be received okay. Although not shown in Figure 2.6, coded PHY in LE adds additional fields to the packet structure. While coded PHY is used, the Access Address is coded via S8, and following the Access Address, a Coding Indicator field (CI) indicates how the rest of the packet is coded (PDU parts: header, payload, MIC, and CRC). The rest of the packet may be coded either as S2 or as S8. The two coding parts algorithm requires trailer terminator patterns that are called TERM1 and TERM2, and are placed after the Access Address (as it is coded as S8) and after the CRC, respectively (as PDU and CRC are coded as S2 or S8). The preamble structure for coded PHY is different, and matches the S8 block coding structure (80 µs long). Also not shown here is an optional Constant Tone Extension (CTE), which was introduced in Bluetooth Core 5.1. This feature enhances Bluetooth's capabilities for direction-finding applications, allowing devices to determine their relative positions with high accuracy.

LE Packet

Preamble	Access Address	Advertising/Data Protocol Data Unit (PDU)	CRC

Advertising Channel PDU

Header	Payload

Data Channel PDU

Header	Payload	MIC

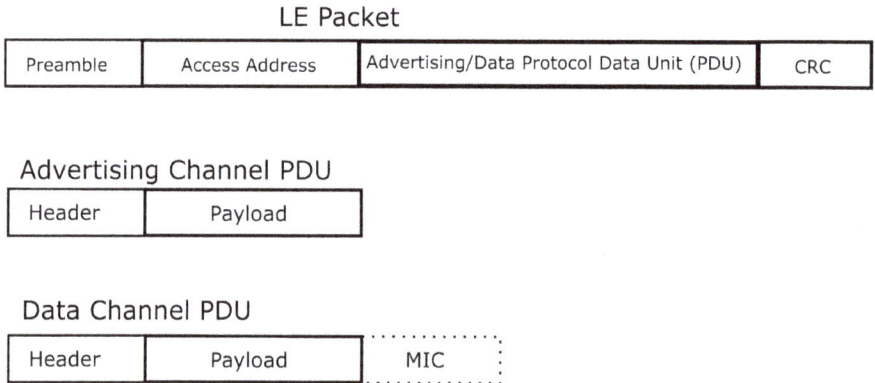

FIGURE 2.6 LE packet format.

HOST AND CONTROLLER

GAP defines four profile roles in Bluetooth® LE: Central, Peripheral, Broadcaster, and Observer.

The Central GAP profile role configures the Controller as the Central device in Bluetooth LE connection (ACL or CIS). The Peripheral GAP profile role configures the Controller as the Peripheral device in Bluetooth LE connection (ACL or CIS). The Central device is the arbiter of the connection and determines when the Peripheral device may access the channel. The Broadcaster GAP profile role configures the Controller as the Bluetooth LE advertiser device or Bluetooth LE BIS Broadcaster. An Observer GAP profile role configures the Controller as the Bluetooth LE scanner device to listen to Advertising data from a Broadcaster device or as the Bluetooth LE Synchronized Receiver to listen for BIS PDUs. The Central may create a Bluetooth LE ACL connection with one or more Peripherals to exchange point-to-point data. While in ACL connection, the Central may initiate a control procedure to create a CIS with each Peripheral it is connected to, to exchange time-bound data. A Broadcaster may advertise data as connectionless to many Observers, which are receiving the Advertising data. A Broadcaster may send data in PA and may create a BIS to broadcast time-bound data to multiple listeners.

Bluetooth® LE uses a similar HCI partition between the Host and the Controller, as in Bluetooth Classic. Bluetooth LE also uses the L2CAP protocol as the basic framing for applications. Above L2CAP, Bluetooth LE deploys the ATT protocol and the GATT profile. The GATT/ATT framework is used by Bluetooth LE applications as a blocking request response protocol, to read and write application data from clients to servers or to notify or indicate changes in data from servers to clients. The GATT profile provides a generic infrastructure to define multiple application services that may be discovered and configured over Bluetooth connection. ATT provides the protocol to read and write or notify GATT characteristics from each application service.

In GAP, the Bluetooth LE Central device is usually a client to a server, which is hosted by a Peripheral device. The Peripheral device is usually hosting the GATT service. An example of a service that is hosted by a Peripheral is a blood sugar moni-tor. A small device attached to a person's arm may read the blood sugar level of a person with diabetes. That device also serves as the Bluetooth LE Peripheral and hosts a blood sugar level monitoring service. The person with diabetes uses a cell-phone application as a Central Bluetooth LE device and as a client to read the GATT service sugar levels or to get an alert notification pushed from the service when the sugar level drops very low. Although it is common for the peripheral device to host the GATT service, there could be cases where the services are hosted by Bluetooth LE Central devices. An example for such a case is a time sync service, where the Central device is publishing network time as a service that a remote peripheral may sync to and get accurate network time. In this case, the Central device may be a PC or a phone with network connectivity to get accurate time, and the remote peripheral or sensor may read the timing adjustments.

GATT and ATT use a fixed L2CAP channel, so multiplexing of applications over L2CAP is not possible. That means that if there are multiple application services over the same Bluetooth LE connection between two devices, then one application may block another. If, for example, one application begins a long read operation, then the other application must wait until the read is completed before initiating a new read. In certain cases where services are considered low-latency and require fast response time, the blocking ATT may be a problem, especially when a few services are sharing the same ACL connection. This problem required enhancing ATT.

Version 5.2 of the Bluetooth Core Specification added an extension to L2CAP and to ATT, which is called EATT (Enhanced ATT). With EATT, it is possible to create a pool of L2CAP channels, so multiple services may issue concurrent read or notify operations, without blocking each other. This is done by configuring small L2CAP frames. The small L2CAP frames allow sending small ATT messages or fragments of long ATT messages, which are later reassembled across the connection. The new protocol enables sending multiple requests or notifications over multiple L2CAP channels. Each L2CAP channel has its own flow control, so one channel does not affect another channel. Later in this book, we will see how LE Audio uses multiple GATT services over the same Bluetooth LE ACL connection between two devices. LE Audio requires EATT to reliably control operations such as stream control, vol-ume control, call control, or media control, which are managed as separate GATT services between two devices. All of these services require a low-latency operation where a non-blocking EATT model is essential for good user experience.

BLUETOOTH® CLASSIC AUDIO

Bluetooth Classic audio is divided into two main use cases – voice and media. The media use case covers music playback, accompanied by other types of media such as movie soundtrack, or system tones from the OS. The voice use case covers phone calls and PC Voice over IP applications (VoIP) for conference calls, video confer-ences, and online sharing meetings. The two use cases (voice and media) are imple-mented by different vertical protocol stacks starting from the application layer down to the Profiles and Protocol and the data link layers. The media/music use case uses an Advanced Audio Distribution Profile (A2DP) with an Asynchronous Connection

(ACL) transport for music data. The Voice use case uses a Hands-Free (HF) profile with the Synchronous Connection Oriented (SCO) or extended Synchronous Connection Oriented (eSCO) transport for voice data.

The difference between SCO and eSCO is mainly in reliability, error detection, and flexibility of packet lengths and scheduling options. Compared with SCO, eSCO provides better reliability. SCO is therefore hardly used today and is considered an outdated technology. The SCO logical transport is a symmetric transport between the Central and Peripheral, where slots are reserved for transmission. Therefore, it is also considered a circuit-switched connection between the Central and Peripheral. The SCO transmission rate is 64 kbps synchronized with the piconet clock. There is no retransmission in SCO.

eSCO is both a symmetric and an asymmetric transport between the Central and Peripheral. Similar to SCO, the slots are reserved for transmission in the case of eSCO, and there are a number of extensions over SCO – flexible packet types, selectable data content, and selectable slot periods. Therefore, eSCO allows for a range of bitrates to be supported over the transport. In addition to reserved slots, eSCO supports retransmission windows immediately following the reserved slots. eSCO is mostly used for voice-only traffic today in Bluetooth Classic. eSCO could have been used for music streaming. However, when eSCO was added to Bluetooth Classic, there was already a profile for music that was using ACL. As a result, eSCO was not utilized for music streaming.

The ACL logical transport is used for best-effort transmission of asynchronous user data. It is also considered to be a packet-switched communication between the Central and Peripheral. Every active Peripheral in a piconet has one ACL logical transport to the piconet Central, known as the default ACL. The piconet Central assigns the default ACL, and it is called the default LT_ADDR (logical transport address). eSCO logical transport is added by the default ACL when a voice session is required, and removed when a voice session ends. eSCO manages packet retries for voice separately from the best-effort ACL. Therefore, eSCO uses a separate LT_ ADDR from the ACL (a secondary LT_ADDR). So packet acknowledgment and sequence numbers are handled separately by ACL and eSCO between two devices. If the default ACL is removed, then all other logical transports that exist between the Central and Peripheral are also removed, such as the eSCO logical transport.

With ACL, as a best effort, a packet may be retried for a longer period of time, while the eSCO transport allows for a very short window of retry, due to the voice delay requirements. The packet types used by eSCO and ACL are different. When Bluetooth® technology evolved, there were not enough bits in the header to describe all possible packet types. There are four bits to describe a packet type, and there are more than 16 packet types in Bluetooth Classic. As a result, the packet type is interpreted differently for eSCO LT_ADDR.

The Hands-Free profile (HFP) uses GSM standard-based AT commands (Attention), for controlling a call and volume levels. The AT commands' purpose is to seek the attention of the modem or call gateway, while remotely invoking operations from a terminal device such as a headset or seeking the attention of the headset when the phone or PC remotely adjusts the headset volume. For the most part, the AT commands are used as is, and they are transported across Bluetooth devices (typically between a phone and a headset or car kit) using a serial port emulation protocol called RFCOMM. The RFCOMM protocol operates over the Logical Link Control and Adaptation Protocol (L2CAP), which provides protocol multiplexing, segmentation and reassembly, and conveying of quality of service information. Examples of AT commands are ATA (for Accepting call), AT+CHUP (for Rejecting call), and so on. HFP may use AT commands to configure CVSD for narrowband speech quality or mSBC for wideband speech quality. Both codecs are configured to run over the eSCO transport.

The A2DP profile uses the Audio/Video Distribution Transport Protocol (AVDTP) for setting up and controlling audio endpoints and the data channels to transport media/music. AVDTP also operates over L2CAP. Along with A2DP, devices typically also implement the Audio/Video Remote Control Profile (AVRCP), which is used to provide remote control functionality (like Play, Pause) and control volume. AVRCP uses another transport protocol called AVCTP (Audio/Video Control Transport Protocol), which again operates over L2CAP. The A2DP profile specifies the mandatory and optional codecs that are used to encode and decode the music stream. It defines SBC (Subband Codec) as the mandatory Codec, and the SBC specification is part of the Bluetooth A2DP profile. SBC uses four or eight subbands, an adaptive bit allocation algorithm, and simple adaptive block PCM quantizers. It inherently supports 16, 32, 44.1, and 48 kHz sampling frequency. And it supports Mono, Dual Channel, Stereo, and Joint Stereo channel modes. Other optional codecs specified by the A2DP profile are MPEG-1, MPEG-2, AAC, ATRAC, and LDAC. The A2DP profile also has a provision for defining vendor-specific codecs.

In the next chapter, we will begin reviewing the LE Audio One Architecture.

REFERENCES

1. Bluetooth Core 6.0 or later, https://www.bluetooth.com/specifications/specs/core-specification-6-0/
2. HFP profile version 1.9 or later, https://www.bluetooth.com/specifications/specs/hands-free-profile/
3. A2DP profile version 1.4 or later, https://www.bluetooth.com/specifications/specs/advanced-audio-distribution-profile-1-4/
4. AVDTP protocol version 1.3 or later, https://www.bluetooth.com/specifications/specs/a-v-distribution-transport-protocol-1-3/
5. AVRCP profile version 1.5 or later, https://www.bluetooth.com/specifications/specs/a-v-remote-control-profile-1-5/

3 One Architecture Overview

This chapter describes a high-level overview of the LE Audio One Architecture. Later chapters will focus on each layer in more detail.

THE LE AUDIO LAYERS

As introduced in Chapter 1, the LE Audio stack forms a One Architecture that is conceptually divided into three main horizontal layers – App, Control, and Transport – and one vertical layer, Codec. The App layer consists of the application and thin use case-based profiles. The Control layer consists of most of the LE Audio host profile stack and deals with audio controls. The Transport layer is part of the Controller and deals with the over-the-air transport of audio data and profile signaling. The App layer uses the Control layer to configure the Transport layer. The vertical Codec layer deals with the data path and interfaces with all of the horizontal layers.

The idea behind One Architecture is to unify all audio-related configuration, control, and audio data, which enables a seamless and power-efficient user experience. The unification enables the Hardware and Software audio vendors to focus efforts on a single architecture. As a result, audio vendors use the One Architecture to innovate and create multiple different use cases in an interoperable way across the audio industry.

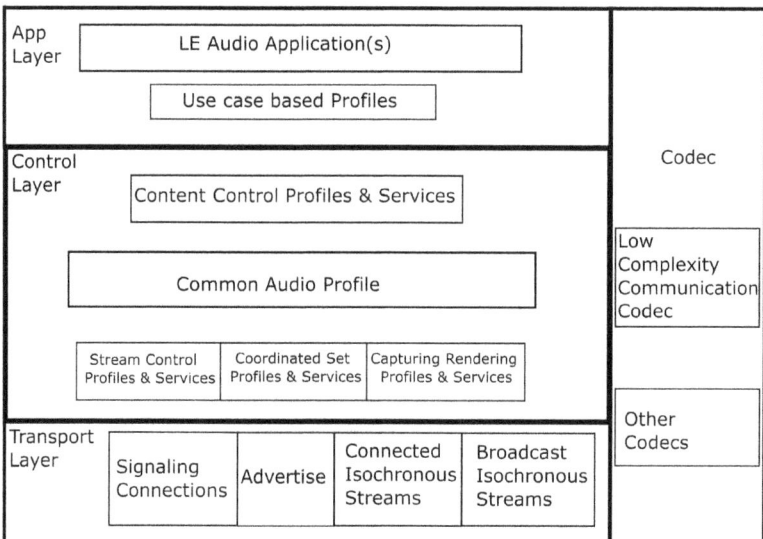

FIGURE 3.1 LE Audio One Architecture.

       DOI: 10.1201/9781003590187-3

Each of the layers of the One Architecture manifests itself as a set of Specification – as either a Host layer protocol, a profile, a service, a Codec Specification, or a Core Specification.

Figure 3.1 shows the LE Audio One Architecture. The App layer is the topmost layer, closer to the end user. The Control layer is the middle layer, which provides all the required controls for the audio setup. The lowest layer is the Transport layer, which is closer to the Bluetooth® LE radio and provides the wireless communication for LE Audio. In the next sections, we will review each layer and its LE Audio components at a high level.

APP LAYER

The application layer consists of the actual end-to-end audio implementation and the use case-based profiles. The application is the primary decision maker; that is, the application has a direct or indirect impact on how the other layers in the architecture function.

The application layer decides and selects the local audio capabilities and the negotiated remote-configured capabilities based on the use case. The application layer configures the audio data path with the selected Codec settings. The audio data path Codec setting decisions translate down to the Transport layer and impact the Core Controller scheduling.

Use case-based profiles add a layer of interoperability to already existing or future possible use cases. Examples of use case-based profiles are the Telephony Media Audio Profile (TMAP), Hearing Access Profile (HAP) and Hearing Access Service (HAS), Gaming Audio Profile (GMAP), and Public Broadcast Profile (PBP), which is also known as Auracast™ broadcast audio. Any applications or manufacturers of devices that adhere to these use cases can implement these use case-based profiles and are practically ready to interop with other devices.

The application layer is designed to be extensible. The definition of new profiles in the application layer serves a use case. And when a new profile is added, it mostly reuses existing components from the Control layer while defining the new use case behavior.

Chapter 4 will focus on each use case-based profile from the application layer in more detail.

CONTROL LAYER

The Control layer consists of multiple different blocks, each serving a different functionality. As described earlier, it is the application that brings individual components together to form a use case. The Control layer is where the LEGO block development model of LE Audio is achieved.

The following sections provide a brief on each of the different blocks and its subcomponents in the Control layer. Chapter 5 will provide more details on each block and each subcomponent in the Control layer.

STREAM CONTROL PROFILES AND SERVICES

Stream control in the context of LE Audio defines the process of configuring and controlling an Audio Stream between two devices, based on the audio capabilities of both devices. The following sections list the primary Specification that forms the Stream Control while providing a brief functionality of each.

Basic Audio Profile

The Basic Audio Profile (BAP) Specification of LE Audio defines the client/server behavior, which forms the basis of all LE Audio Stream control. BAP describes the basic roles that are required from implementations to define a Unicast or Broadcast behavior. BAP specifies how the servers need to interact with the client and defines the message sequencing between the client and the server. BAP defines the device discovery, connection establishment, Advertising, and scan procedures used by the client, server, Broadcast source, and Broadcast sink. BAP uses GATT control point operations for command control. BAP specifies the synchronization between multiple servers in order to synchronize the timing of audio playback and audio capture.

Unicast

BAP Unicast audio deals with setting up point-to-point streams between a client and a server. In Unicast topology, the clients implement the Bluetooth® LE GAP Central role and are typically implemented by devices such as PCs or phones. Servers implement the Bluetooth LE GAP Peripheral role and are typically implemented by speakers, headsets, wireless microphones, or voice assistant systems. In the link layer, the BAP Unicast client is the Bluetooth LE Central device, and the BAP Unicast server is the Bluetooth LE Peripheral device.

A Unicast stream in BAP is always configured to flow in one out of two possible directions – either from a client to a server or from a server to a client. For a voice call, two streams in opposite directions are required between a client and a server – one sink stream and one source stream. In Unicast topology, the descriptor of the stream direction sink or source is always relative to the server device (the Bluetooth LE Peripheral), which is also known as the Unicast endpoint. So, for example, when answering a simple phone call from an earbud, two streams are enabled: one sink stream to the earbud speaker and one source stream from the earbud microphone. In the case of a voice call with two BAP streams to the same device, a single bidirectional transport is used to map to the BAP sink stream and BAP source stream.

An application or use case-based profile may use a BAP Client to set up streams to more than one server in order to send Unicast audio from one client to multiple server devices. On the other hand, a BAP Server device may enable more than one client to stream audio by exposing a set of audio endpoints, either concurrently or based on a time-sharing basis. This concept is also known as the sink/source-led journey, which means that a user of an LE Audio client or server may have the flexibility to select which sources of audio to render or capture to and from the terminal devices.

Broadcast

BAP Broadcast Audio deals with setting up one-to-many connectionless Broadcast streams. It defines the roles of the Broadcast source and the Broadcast sink. A Broadcast source may transmit to an unlimited number of Broadcast sinks. Examples for a Broadcast source may be PCs, phones, TVs, or public announcement systems. Examples of Broadcast sinks are headsets, hearing aids, or a set of bookshelf speakers (two or above). A Broadcast source implements the Bluetooth LE GAP Broadcaster role, as it advertises and streams the audio, with no specific target device. A Broadcast sink implements the Bluetooth LE GAP Observer role, since it synchronizes to Periodic Advertisements from the Broadcast source and to the Broadcast Audio Stream.

A Broadcast stream in BAP is always configured to flow in a single direction and always from the Broadcast source to the Broadcast sink. In Broadcast topology, the descriptor of the stream direction is relative to the source device, which is also known as the Broadcast source endpoint. The Broadcast sink is required to synchronize and adapt to the Broadcast source endpoint configuration.

In the link layer, the Broadcast source is broadcasting connectionless stream data and uses Periodic Advertisements to publish the stream control, Codec, QoS (quality of service) settings, and stream metadata about the context of the audio use case. In the link layer, the Broadcast sink listens to the Periodic Advertisements from the Broadcast source to learn about the stream content and timing, and synchronizes to the Broadcasted stream in order to consume and render the audio content.

The Broadcast sink may scan in order to synchronize to the Periodic Advertisements, or it may receive the synchronization information directly from another device or from the Broadcast source. A link layer protocol exists, which is called PAST (Periodic Advertising Sync Transfer). With PAST, it is possible to send the timing information of the Periodic Advertisements over a Central/Peripheral Bluetooth LE connection. This way, Broadcast sinks do not need to scan in order to synchronize to a Broadcast stream and therefore save more power.

Endpoints

In Unicast audio, the endpoint is called ASE (Audio Stream Endpoint). ASEs are records that exist on a server device. There are two types of ASE records: Sink ASE and Source ASE. The ASE configuration includes Codec settings, such as sampling rate, bitrate, SDU (Service Data Unit) size, and QoS parameters for interval, retry effort, and latency. ASE records are offered by servers and configured by clients.

In Broadcast audio, the endpoint is called BASE (Broadcast Audio Source Endpoint). The BASE is always in one direction – source – and always relative to the Broadcast source device. That means, for example, that unlike in Unicast audio, the endpoints in Broadcast audio do not exist at the headset device. In Broadcast audio, the headset discovers the Broadcast source endpoint, while in Unicast audio, a PC, for example, discovers endpoints in the headset. This is a subtle behavior that you should remember while we go back and describe Broadcast in more detail in later chapters.

The BASE information is advertised in Periodic Advertisements, as Host Advertising data. The Periodic Advertisements also contain link layer information called BIGInfo, which describes the timing of the actual Broadcast stream. The BASE, however, describes the Codec settings and context use case info about what audio content is carried over the Broadcast. A receiver of a Broadcast needs to know the Codec settings in order to decompress the audio content and the context about what audio is published in the Broadcast stream, such as the TV program name or language. In certain cases, a Broadcast stream may be encrypted, and this info is described in the BASE and in the BIGInfo.

Published Audio Capabilities Service

The Published Audio Capabilities Service (PACS) is a GATT-based service that is used by Unicast servers and Broadcast sinks to expose their audio capabilities. Audio capabilities are defined in the form of a PAC record and are separately exposed for the server's Audio Source or Audio Sink. Audio capabilities also include the Audio Location supported by the device (e.g., Left, Right) – which is used to distinguish and treat that device accordingly (e.g., the left earbud and right earbud will receive different content for a stereo playback). PACS also exposes the context availability of the device to provide the client with information relating to what content can be supported by the device and what content it is currently engaged in. Context and content in this case refer to high-level use cases – like a phone call or music playing. PAC records are used in both Unicast topology and Broadcast topology. In Unicast topology, the client uses the PAC records from the server when performing ASE configuration. In Broadcast topology, when scanning for Broadcast sources is done by a third device, the PAC records are used to determine if the Broadcast sink has the capabilities to synchronize to the BASE (e.g., a phone configures earbuds to synchronize to a Broadcast source).

Audio capabilities of a given server may be cached by a paired client. For example, a PC can remember that an LE Audio headset with Bluetooth® address AA:BB:CC:DD:EE:FF has the capability to stream stereo to its left and right speakers and to capture mono source audio from its microphone. The PC, for example, may also read and cache PAC records about the specific supported Codec settings with sampling rates of 32 kHz and 48 kHz, and the supported Codec bitrates and SDU frame interval. This, in turn, enables faster Audio Stream establishment as the process of audio capability discovery may be skipped in subsequent connections to the same Bluetooth address.

Audio Stream Control Service

The Audio Stream Control Service (ASCS) is a GATT-based service and is used by servers to expose an interface for ASEs for Unicast audio. ASE enables clients to discover, configure, establish, and control context over the Unicast Audio Streams. A Unicast Audio Stream is a logical communication channel between the client and the server.

ASCS defines the basic ASE state machines and their transitions. The basic states are "Idle," "Codec Configured," "QoS Configured," "Enabling," "Streaming," "Disabling," and "Releasing." The basic operations that are applicable to these state machines are "Config Codec," "Config QoS," "Enable," "Disable," and "Release." The transition of an ASE from one state to another is based on these operations. There are other operations on the ASEs such as "Receiver Start Ready," which controls when Audio Context is sent in the Audio Stream, and "Update Metadata," which controls what Audio Context is sent in the Audio Stream. An App layer profile controls when to start streaming audio, in what context, and may update the streamed Audio Context according to the use case.

The ASCS service is the interface between the Host profiles and the Core Isochronous Channels, which is the LE Audio data transport. When ASE is in the "Streaming" state, it is linked to an underlying link layer CIS transport. A brief explanation about CIS is provided later in this chapter and in more detail in Chapter 7.

The CIS transport parameters are governed by the QoS parameters, which are specified by the client during QoS configuration. ASE details may be cached, depending on the caching policy of the client and server, which allows the ASE configuration to be reused across connections.

Broadcast Audio Scan Service

The BASS is a GATT-based service offered by audio peripherals to delegate scanning to a third scan assistant device. BASS is used for scan assistance and encrypted Broadcast key installation. Scan assistance is a technique used by battery-constrained devices (like earbuds) to delegate scanning for a Broadcast source to a third assisting device. A typical example is a user holding a phone to scan and select Broadcast sources and passing this information to the earbuds to synchronize to that Broadcast source.

The BASS service allows an assisting client to configure the server to what Broadcast source to listen to. The assisting client will use information from the server PAC records to determine if the server has matching capabilities to synchronize with the Broadcast source. The assisting client uses a link layer procedure to pass the timing information about the PA. The client can request the server to synchronize to the PA and other relevant information and synchronize to the Broadcast stream in the link layer, which is called BIS. A brief explanation about BIS is provided later in this chapter and in more detail in Chapter 7.

A scan assistant can pass the Broadcast code for the decryption of an encrypted audio Broadcast.

PACS and ASCS together are required to define the Unicast server behavior. PACS and BASS together are required to define the Broadcast server behavior.

CAPTURING AND RENDERING PROFILES AND SERVICES

Capturing and rendering in the context of LE Audio are defined as the process of controlling the speaker or microphone gain and mute settings. The following sections describe the primary blocks that constitute the Capturing and Rendering profiles and services and highlight the LE Audio Specification that implements this functionality. Chapter 5 provides more details.

Volume Control

Volume control use cases are implemented using the GATT-based Volume Control Profile (VCP) and Volume Control Service (VCS). The VCS is implemented on devices that control the volume of an audio output device, such as a speaker. VCP defines procedures to perform relative as well as absolute volume control. It also defines procedures to control the mute state of the speaker.

VCS may optionally include the following secondary services:

VOCS (Volume Offset Control Service) may be used when a server device has multiple speakers and allows control of each of the speakers' volume levels individually. In this case, in addition to VCS, which controls the main volume, each of the speakers may be allowed to offset the volume level such that certain speakers may be higher than others. An example is a balance between a left and a right speaker in a car kit speaker system or equalizing volume levels among multichannel settings with front, rear, and center speakers.

AICS (Audio Input Control Service) may be used if the server device allows controlling switching off/on the volume of different audio sources, which may be connected to the device speakers. An example is a hearing aid device that may receive audio over Bluetooth® technology from a phone and may also receive an ambient sound, which is amplified by a local microphone. People with hearing loss use the hearing aid to amplify the sound in their surroundings. With AICS, the volume control may be selected to act on Bluetooth stream only or on ambient sound only, or on both.

Microphone Control

Microphone control use cases are implemented using the GATT-based Microphone Control Profile (MICP) and Microphone Control Service (MICS). These are used to control the mute and gain properties of the microphone when capturing or recording audio. The MICS is declared on devices with control over microphone audio.

MICS may also include AICS to control input from multiple microphones in the same device, such as a microphone array.

CONTENT CONTROL PROFILES AND SERVICES

Content control in the context of LE Audio is defined as the control of content, such as telephony control or media control. The following sections describe the primary blocks that constitute the Content control profiles and services, and also highlight the LE Audio Specification that implements this functionality. Chapter 5 contains more details.

Call Control

The Call control use case provides voice call content and context types. Call Control is implemented using the GATT-based Call Control Profile (CCP) and Telephony

Bearer Service (TBS). The TBS provides the call gateway service and is implemented on devices that can make and receive phone calls, such as phones or laptops or PCs. This is an example of a service that is hosted by a Central device. In the case of TBS, the device that is connected to the network, to provide cellular access or voice over IP access, is the Central device. The CCP defines roles and procedures to remotely interact with devices that implement the TBS. The CCP client is implemented by devices such as headsets with buttons or voice commands to answer calls or terminate calls, and in devices such as car kits with displays to show information about an ongoing call.

There are sets of procedures defined in CCP to initiate call control operations and to determine the information related to the bearer (e.g., Cellular, VoIP). The Call control operations are performed using a write to a GATT control point and are related to the basic call control functionality, like answer call, terminate call, place call, join call, and call hold options. The information retrieval procedures are performed using reading and notifying GATT characteristics.

The TBS Specification is designed so that multiple call control bearers may coexist. It defines a Generic Telephone Bearer Service (GTBS) as well as a specific Telephone Bearer Service (TBS). GTBS aggregates and unifies its internal telephone bearers into a single bearer and provides the telephone status and control of the device as a single unit. The device maps the status and control to its internal telephone bearers in an implementation-specific manner. GTBS is suited for lightweight clients that do not need to access specific telephone bearers. The behavior of TBS and GTBS is identical.

Media Control

The Media control use case provides media playback content and context types. Media Control is implemented using the GATT-based Media Control Profile (MCP) and Media Control Service (MCS). The MCS is implemented on devices that have the actual media player, such as phones, music players, or laptops/PCs. The MCP defines the roles and procedures to remotely interact with devices that implement the MCS. The MCP client is implemented by devices such as headsets with buttons or voice commands to send play or pause commands and in devices such as car kits with displays to show information about the played track and album, such as the song artist, title, and an album cover picture.

MCP defines the procedures related to media playback and control, like exchanging track information, controlling the current track (play, pause, etc.), moving back and forth inside the track as well as within tracks in an album, and changing playback speed. MCP is optionally dependent on the Object Transfer Profile (OTP) and service for exchanging large metadata information regarding the track that is playing (such as pictures).

The MCS Specification, just like the TBS Specification, is designed to support multiple media players simultaneously. It defines a Generic Media Control Service (GMCS) as well as a specific MCS. The GMCS provides the status and control of the media playback for the device as a single unit, and it is used by lightweight clients. An MCS instance describes and controls the media playback for a specific media player within the device. A device implements MCS instances to allow clients to access the separate internal media player entity; multiple MCS instances may be used by clients with rich user interfaces and displays. The behavior of MCS and GMCS is identical.

COORDINATION SET PROFILES AND SERVICES

With coordination control, a set of devices may be discovered and treated as a single set, for example, use cases like discovering left and right earbuds of the same set or multiple speaker pieces of a multichannel surround system.

Coordination is implemented by the GATT-based Coordinated Set Identification Profile (CSIP) and Coordinated Set Identification Service (CSIS). A collection of audio peripherals that belong to the same set is considered a set member.

Each set member implements the CSIS as a GATT service. A client device, such as a phone or PC, implements CSIP and may discover a set by connecting to one set member, reading token information from CSIS, along with the size of the set (how many set members). The token information is called the Set Identification Resolution Key (SIRK). Once the SIRK is retrieved, the client can use it to scan for all other set members with a similar hashed token, which is advertised as a Resolvable Set Identifier (RSI). Additional authentication is done after all set members are connected to the client. This procedure allows a client to quickly discover and connect to a set by filtering out addresses that are not part of the lookout set.

COMMON AUDIO PROFILE

The Common Audio Profile (CAP) deals with synchronization across multiple devices. It specifies procedures to start, update, and stop Unicast and Broadcast Audio Streams on groups of devices. It also specifies procedures to control volume and microphone input on groups of devices. It uses the procedures from the underlying BAP, VCP, MICP, and CSIP profiles.

CAP defines the association of Context Type values with the Unicast and Broadcast Stream. Context Type values specify the type of content that is transmitted by the Audio Stream. The Context Type values are defined in BAP, but CAP specifies the usage. Multiple types of context may be linked to a single stream.

CAP defines the association of Content Control Identifiers (CCIDs), which identify a specific instance of a Content Control Service (either TBS or MCS) on the server. Since there could be multiple instances of TBS and MCS services on the server, the CCID is used to distinguish and uniquely identify one instance. Multiple CCIDs may be linked to a single stream as a CCID list.

This Specification includes the Common Audio Service (CAS), which is used to discover basic CAP-related feature support on the server and enables the coordinated audio setup of multiple peripherals by CAP.

CODEC LAYER

The Codec layer mandates the use of a single Codec by all LE Audio–compliant devices while allowing the use of optional codecs or vendor-specific codecs. Mandating a single Codec guarantees a basic level of interoperability. The other codecs address specific use cases such as multichannel surround content. The following section will provide an introduction to the mandatory LE Audio Codec: LC3 (Low Complexity Communication Codec). Chapter 6 provides more details about LC3.

Low Complexity Communication Codec

LC3 is the single mandatory Codec in LE Audio. When devices from different use cases are engaged for audio, LC3 allows for a lower common denominator, and that audio may be compressed and sent over the Bluetooth® transport, knowing that the remote side has the capability to decompress the audio packets.

LC3 operates using a transformation on a sampling frame buffer. Audio PCM samples are aggregated into a frame buffer prior to the transformation and compression steps. The number of samples to compress depends on the sampling rate. As we saw in Chapter 1, voice audio requires a lower sampling rate compared with music audio, since music audio produces higher tones, which require sampling higher audio frequencies.

For example, LC3 for high-quality music requires a PCM sampling rate of 48 kHz. The length of the sample buffer frame is measured in milliseconds (ms). LC3 supports two sample buffer frame durations: 7.5 ms and 10 ms. Music sampled at 48 kHz generates 480 samples every 10 ms or 360 samples every 7.5 ms. LC3 allows configuration of the compression ratio. For example, sampling of music at 48 kHz may compress audio, such that it results in a few options of bitrates. The bitrate is a parameter of LC3. The profiles for high-quality music may use bitrates of 320 kbps, 160 kbps, 124 kbps, 96 kbps, and 80 kbps. Bluetooth SIG listening tests show excellent audio quality with a bitrate of 96 kbps for the music in the left or right channel. The PCM sample buffer also depends on the sample depth. LC3 supports the following bit depths: 8, 16, 24, 32 (note that the internal quantizers of LC3 are 16 bits and 24 or 32 bit depths are supported as input/output interface only). The common bitrate configuration for LC3 is with 16-bit sample depth, which got a high score in listening tests done by the Bluetooth SIG.

The combination of bitrate and frame duration defines the size of the SDU, which is sent in every Codec frame interval. One instance of the LC3 Codec generates one SDU every frame duration, and this SDU is sent for transmission over the LE Audio transport. For example, LC3 for music at 48 kHz with 96 kbps bitrate generates SDUs at a size of 90 octets for a frame duration of 7.5 ms or SDUs of 120 octets for a frame duration of 10 ms. (Sample calculation for 10 ms frame duration: [1000 ms/10 ms] × [120 octets × 8] = 96000 bps = 96 kbps.)

When an application configures a use case, it is therefore required to know in advance the supported Codec capabilities and how the Codec may be configured in order to provide the required sound quality per the selected use case. Based on the Codec settings, a set of QoS settings is determined for the LE Audio transport layer. In addition to bitrate, there are other QoS settings such as maximum allowed latency and retry reliability configuration.

Audio Latency and Retry Effort

The LC3 Codec creates self-latency during audio creation and compression. A full PCM sample buffer is required prior to preparing a single SDU for transmission. It means that at the encoder side, at least a frame duration worth of latency is already accumulated. This figure is either 7.5 ms or 10 ms, which depends on the frame duration configuration. In addition to this sampling delay, the Codec requires an additional inherent algorithmic delay. On the receiver side, similar decoding and buffering delays are required during the decompression of the received SDU into PCM. And the wireless transmission over Bluetooth® also adds delay. The various net delay factors add up to about 20 ms when the frame interval is 10 ms or to about 17 ms when the frame interval is 7.5 ms.

For voice applications, the QoS requires as low latency as possible, so minimum latency is configured based on what LC3 may provide. That means that additional latency should be limited in the over-the-air transmission, and all retries of the coded SDU must complete before the next LC3 frame interval. The same is true for media applications such as movie audio, where low-latency audio is required to sync the moving picture and the soundtrack audio, so the eyes and ears get synchronized content. This is also known as Audio/Video or AV sync. Low latency for AV sync may not be as small as for voice, however, as it is possible to buffer video and synchronize to audio rendering.

There are other applications, such as listening to recorded music. In this case, the media player allows for larger delays up to 100 ms, since the application player or renderer speaker device may buffer audio (also known as a jitter buffer) and compensate for delayed frames by varying the playback speed to compensate for transmission delays. In these cases, there is more freedom to allow a larger number of retries, such that SDU from the LC3 Codec may be retried for multiple intervals over the air. On top of the end-to-end latency, a presentation delay at the server is added to manage local delay at the server from the reception of SDU to actual playback on the speaker. Some examples of Presentation delays may be as low as 5 ms or as high as 60 ms, depending on the specific audio equipment.

Presentation delays are exposed by the server and configured by BAP.

TRANSPORT LAYER

The Transport layer consists of two different types of transport for audio – Unicast and Broadcast – and two transports for signaling – Advertising and signaling connection. The following sections provide a brief overview of each of the four transports used in LE Audio. Chapter 7 provides more details on how these transports are used in LE Audio.

UNICAST TRANSPORT – CONNECTED ISOCHRONOUS STREAM

CIS enables a point-to-point unidirectional or bidirectional audio flow between two devices, a Central and a Peripheral. A collection of CIS transports forms a CIG (Connected Isochronous Group). A central may establish multiple Audio Streams to a few peripherals so they are grouped together as a single use case.

CIS is established via a link layer control procedure and contains a collection of QoS parameters to determine bitrate, retry effort, and transport latency. CIS is considered a time-bound transport and manages synchronization across multiple CIS in a CIG, such that a synchronization reference point is maintained across multiple Audio Streams. Data PDUs in a CIS have a lifetime that is determined by the retry effort and the transport latency. At the scheduler, these parameters translate to sending data at fixed intervals with a retry window and flush timeout. CIS introduces subevents within each CIS event. The subevents allow sending different parts of an audio SDU or a retry of audio data at different RF frequencies (and therefore improve link quality).

BROADCAST TRANSPORT – BROADCAST ISOCHRONOUS STREAM

BIS enables a point-to-multipoint unidirectional connectionless audio flow between one source device, a Broadcast source, and many Broadcast sinks. A collection of BIS transports forms a BIG (Broadcast Isochronous Group).

BIS is established autonomously in the source, and its timing and configuration details are advertised in PA trains. BIS contains QoS parameters to control the number of unconditional retries and latency. Data PDUs in a BIS are sent in multiple copies at fixed intervals and within each interval. BIS introduces subevents in each event, which allow sending different parts of audio or different copies of audio PDUs. BIS contains reference timing info, which is advertised, and it enables synchronized receivers to receive audio in unison from one BIS or multiple BIS in a BIG.

ADVERTISING

Advertising enables Bluetooth® LE devices to publish pre-connection info in the case of CIS and Broadcast source info in the case of BIS. In the case of CIS, the pre-connection info helps to determine the availability of devices to begin audio use cases. In the case of BIS, PA provides a means for a Broadcast source to publish the source endpoint audio status, settings, and stream timings such that Broadcast sinks may synchronize and begin consuming content.

SIGNALING CONNECTION – ACL

The signaling ACL connection enables the exchange of Control layer profile data between LE Audio clients and LE Audio servers. All GATT profile data is sent over the ACL transport as L2CAP packets. In the case of CIS, ACL is used to form CIS via link layer control PDUs. ACL is used in both Unicast audio and Broadcast audio. In Unicast audio, clients and servers exchange GATT info over ACL. In Broadcast, clients use ACL to send GATT discovery and configuration information and link layer timing information when a scan assistant configures a peripheral device to synchronize to a Broadcast source.

LE AUDIO HOST AND CONTROLLER

While we have reviewed the LE Audio One Architecture, it is also essential to understand the functional split based on the Bluetooth® Host and Controller in the context of LE Audio.

Figure 3.2 shows the LE Audio split between the Host and Controller parts. The LE Audio host part consists of the profiles and upper core signaling protocols via L2CAP and GATT. The Controller part consists of the transport layer audio path (CIS and BIS) and the signaling via L2CAP ACL Data PDUs, link layer control PDUs, and Advertising PDUs.

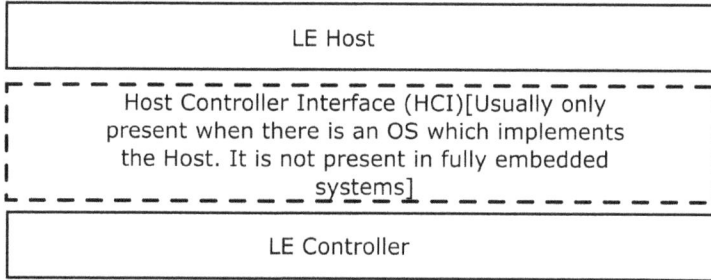

```
┌─────────────────────────────────────────────────────────┐
│                                                         │
│                        LE Host                          │
│                                                         │
├─ ─ ─ ─ ─ ─ ─ ─ ─ ─ ─ ─ ─ ─ ─ ─ ─ ─ ─ ─ ─ ─ ─ ─ ─ ─ ─ ┤
│            Host Controller Interface (HCI)[Usually only  │
│          present when there is an OS which implements    │
│            the Host. It is not present in fully embedded │
│                        systems]                          │
├─ ─ ─ ─ ─ ─ ─ ─ ─ ─ ─ ─ ─ ─ ─ ─ ─ ─ ─ ─ ─ ─ ─ ─ ─ ─ ─ ┤
│                                                         │
│                     LE Controller                       │
│                                                         │
└─────────────────────────────────────────────────────────┘
```

FIGURE 3.2   Bluetooth® LE Host and Controller.

The Codec may reside in either the Host part or the Controller part, and the choice is implementation-specific. The choice of placement of the Codec follows the implementation choice of the audio path between the Host and the Controller. Routing of audio samples to and from the Host CPU and from and to the Controller may use a dedicated PCM bus. In this case, the Codec may reside in the Controller. Alternatively, the Host may compress or decompress the samples and send or receive compressed audio packets to and from the Controller.

The Host/Controller split is common in devices that follow the operating system model. These devices are typically PCs and phones, although it is becoming common today for speakers and other peripheral devices, such as headsets, to adopt the operating system model. In the operating system model, the Controller is a different board, with a bus interface to a separate CPU subsystem, which runs the Host part.

On the other hand, there are peripheral devices that adopt the fully embedded model without the Host. These peripherals may be headsets, earbuds, or wireless microphones. In this case, the profiles, Codec, signaling, and transports are all implemented within a single controller, together with the speakers and microphones' physical parts. Also, in the fully embedded model, it is common to use the terms Host and Controller, although there is no real physical split. In Bluetooth technology, the Host part is considered anything above the HCI functional partition. HCI may use a physical bus to convey packaged messages between the Host and the Controller, or it may be a logical function interface, with no physical bus in cases such as a fully embedded model.

The LE Audio host stack separates the control path from the data path. The control path is used to establish a logical communication Audio Stream. And the data path enables data to flow in this established Audio Stream. The scheduler in the Controller uses control signaling to signal and realize the required Host QoS contracts per the use case. The scheduler also applies implementation-specific decisions on how and when to send and receive audio data, given all other Bluetooth technology-related activities in the system. For example, a PC streaming over LE Audio may also need to communicate with a Bluetooth® LE mouse and a Bluetooth LE keyboard. Or a phone streaming LE Audio may also simultaneously communicate to a watch, a heart rate monitor, a sugar level monitor, or a pedometer in sports shoes.

HOST

The Host stack for LE Audio is primarily based on the GATT client service model. The LE Audio Control layer defines a collection of GATT services and a set of LE Audio profiles. The LE Audio profile defines the behavior and relationship between clients and servers. Clients and servers are logical entities, and each Bluetooth® LE device may manage multiple clients and multiple servers in order to realize one use case for the App layer. This is a key feature of the LE Audio One Architecture. The architecture is divided into smaller client–server interactions. Each of these client–server interactions serves a dedicated purpose or a control aspect of audio. This partition allows for the encapsulation of dedicated functionality in multiple small profiles, instead of one large monolithic profile. An audio use case typically uses a subset of client–server interactions to define the desired behavior.

EATT

LE Audio leads to one of the essential enhancements to the Core Bluetooth® LE Host stack – Enhanced Attribute Protocol (EATT) and Enhanced L2CAP, which allows concurrent transactions to be handled by the stack. This is different from ATT (Attribute Protocol), which leads to blocking of Attribute transactions while one transaction is ongoing.

The various services provide multiple controls such as stream discovery, stream establishment, call, media, recognition, volume, microphone, routing, and coordination. All controls may run in parallel. In classic ATT, one control interaction blocks the other, and there was a need to create an enhancement to provide parallel transactions without blocking each other.

EATT is designed to circumvent the limitation of blocking calls with the classic ATT protocol – which is the very basic protocol for the GATT profile. EATT transactions are designed to be executed in parallel if they are supported by distinct L2CAP channels that use the Enhanced Credit-Based Flow Control Mode, that is, distinct Enhanced ATT Bearers. EATT uses smaller frame sizes over L2CAP channels so that a larger GATT message on one channel will not block a smaller GATT message on a different channel.

CONTROLLER

The Core Controller Stack defines the Core Connected Isochronous Stream and Group (CIS and CIG) and Broadcast Isochronous Stream and Group (BIS and BIG). CIS and CIG are connection-oriented communication and would be used for Unicast use cases. BIS and BIG are connectionless Isochronous communications and would be used for Broadcast use cases.

The HCI layer is enhanced by a number of new commands and events to allow for Isochronous communication to be configured and used.

The Isochronous Adaptation Layer (ISOAL) is provided on top of CIS and BIS to allow for audio SDU framing, similar to how L2CAP packets are framed over ACL. ISOAL supports segmentation and reassembly of different sizes of Host layer data packets, which may differ from Core Controller CIS and BIS configured data packet sizes (PDUs). ISOAL contains synchronization rules for data transmission and reception over CIS and BIS transports, via reference point definitions.

REFERENCES

1. Bluetooth Core 6.0 or later, https://www.bluetooth.com/specifications/specs/core-specification-6-0/
2. BAP version 1.0.2 or later, https://www.bluetooth.com/specifications/specs/basic-audio-profile-1-0-2/
3. PACS version 1.02 or later, https://www.bluetooth.com/specifications/specs/published-audio-capabilities-service-1-0-2/
4. ASCS version 1.0.1 or later, https://www.bluetooth.com/specifications/specs/audio-stream-control-service-1-0-1/
5. BASS version 1.0 or later, https://www.bluetooth.com/specifications/specs/broadcast-audio-scan-service/
6. VCP version 1.0 or later, https://www.bluetooth.com/specifications/specs/volume-control-profile-1-0/
7. VCS version 1.0.1 or later, https://www.bluetooth.com/specifications/specs/volume-control-service-1-0-1/
8. VOCS version 1.0.1 or later, https://www.bluetooth.com/specifications/specs/volume-offset-control-service-1-0-1/
9. AICS version 1.0 or later, https://www.bluetooth.com/specifications/specs/audio-input-control-service-1-0/
10. MICP version 1.0 or later, https://www.bluetooth.com/specifications/specs/microphone-control-profile-1-0/
11. MICS version 1.0 or later, https://www.bluetooth.com/specifications/specs/microphone-control-service-1-0/
12. CCP version 1.0 or later, https://www.bluetooth.com/specifications/specs/call-control-profile-1-0/
13. TBS version 1.0 or later, https://www.bluetooth.com/specifications/specs/telephone-bearer-service-1-0/
14. MCP version 1.0 or later, https://www.bluetooth.com/specifications/specs/media-control-profile/
15. MCS version 1.0.1 or later, https://www.bluetooth.com/specifications/specs/media-control-service-1-0-1/
16. CSIP version 1.0.1 or later, https://www.bluetooth.com/specifications/specs/coordinated-set-identification-profile-1-0-1/
17. CSIS version 1.0.1 or later, https://www.bluetooth.com/specifications/specs/coordinated-set-identification-service-1-0-1/
18. CAP version 1.0 or later, https://www.bluetooth.com/specifications/specs/common-audio-profile-1-0/
19. CAS version 1.0 or later, https://www.bluetooth.com/specifications/specs/common-audio-service-1-0/
20. LC3 version 1.0.1 or later, https://www.bluetooth.com/specifications/specs/low-complexity-communication-codec-1-0-1/

4 App Layer

In the previous chapter, we reviewed the LE Audio stack at a high level. Starting with this chapter, we will begin focusing on each layer in more detail.

We will start from the App layer and slowly move down the LE Audio stack.

In this chapter, we will study the LEGO block model of the application development of LE Audio.

Later chapters will focus on the Control, Codec, and Transport.

ARCHITECTURE

Figure 4.1 describes how the LE Audio Architecture of the App layer fits with the overall LE Audio One Architecture. The App layer contains the LE Audio applications and the use case-based profile modules which serve a collection of different use cases.

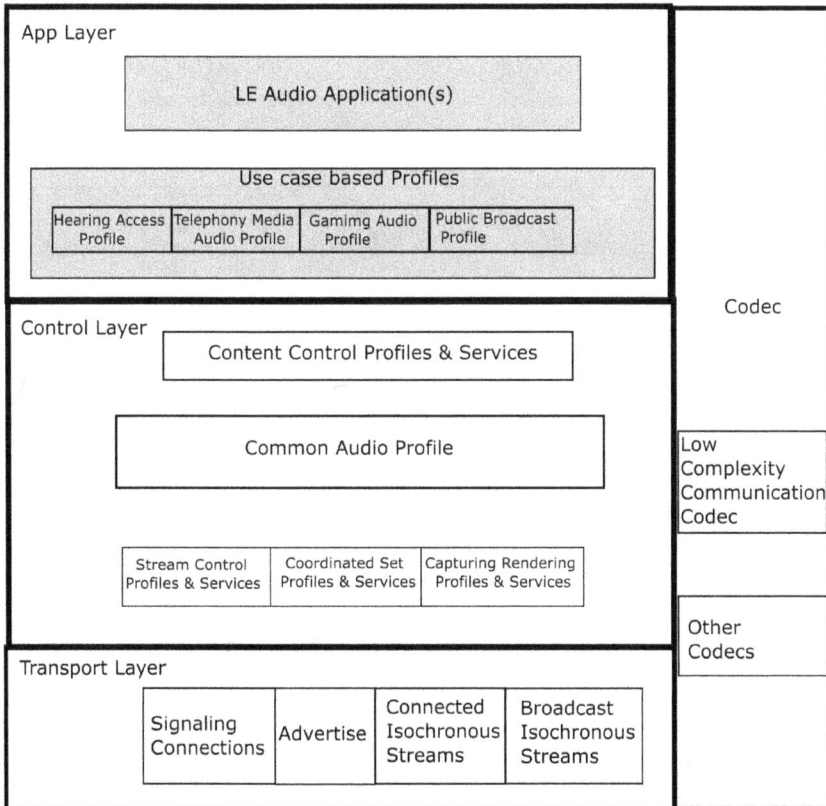

FIGURE 4.1 LE Audio Application layer modules [highlighted boxes].

DOI: 10.1201/9781003590187-4

Table 4.1 lists the different key applications and their relationship to use case-based profiles and Control layer components shown in Figure 4.1. The LE Audio application is scalable by having a common functionality encapsulated in a single layer – CAP from the Control layer. All LE Audio application profiles follow CAP when accessing the lower building blocks. This way, the architecture supports emerging future LE Audio applications, which will be interoperable by using the same CAP rules.

TABLE 4.1
Application to LEGO Layer Mapping

Use case	LEGO Layers	Description
Hearing aid	Common Audio Profile (CAP) Content control (call, media) Hearing Access Profile (HAP) Hearing aid application	Setting up audio for people with hearing loss
Telephony	Common Audio Profile (CAP) Content control (call) Telephony and Media Audio Profile (TMAP) Telephone application	Setting up mono voice calls
Media	Common Audio Profile (CAP) Content control (media) Telephony and Media Audio Profile (TMAP) Media player application	Setting up high-quality media playback for music, movies, or gaming, with both media and voice
Gaming	Common Audio Profile (CAP) Content control (media) Content control (call) Gaming Audio Profile (GMAP) Gaming and low-latency applications	Setting up gaming applications in which both media playback and voice are mixed. The playback is stereo, and the voice may be mono or stereo
TV	Common Audio Profile (CAP) Public Broadcast Profile (PBP) Public TV application	Setting up Broadcast Audio for TV or music in public spaces

LE AUDIO APPLICATION

LE Audio Application is the implementation-specific component of the application stack that brings all the relevant LE Audio blocks together and constructs the user experience. It is responsible for ensuring that the applicable procedure of the right LE Audio block is invoked at the appropriate time for enabling any use case. It is also responsible for providing a rich user interface based on the form factor of the device for ease of use. It is possible that most of the LE Audio blocks get implemented by an operating system, and the application, in that case, has to use the appropriate APIs provided by the operating system. There may be cases where the application needs to include libraries (both static and dynamic) to include the functionality of the LE Audio. All of that is specified by the API interface of the LE Audio blocks.

The application and the API interface of the LE Audio blocks are all implementation-specific, and Bluetooth® SIG does not define that in the Specification.

USE CASE-BASED PROFILES

Use case-based profiles are designed to keep in mind the end-user-visible use case. These aid the application developer in rapidly developing the use case. These profiles take care of the heavy lifting of organizing the relevant Profiles and Services required for the specific use case. They rely on the underlying CAP and Content Control Profiles and Services.

HEARING ACCESS PROFILE

The Hearing Access Profile (HAP) is designed for developing an interoperable LE Audio-based hearing aid device compatible with the rest of the hearing aid ecosystem. It also mandates minimum product requirements for devices that will be based on this Profile. It specifies certain mandatory and recommended configurations to be supported by the lower layers.

Table 4.2 describes the HAP roles and functionality.

TABLE 4.2
HAP Roles and Functionality

Role	Description of Functionality	Examples
HA	Hearing Aid Role. It mandates the following functionality: Decode of at least one Standard Quality and one High-Quality audio channel. The ability to accept the establishment of both Unicast and Broadcast Audio Streams. Support for both Bluetooth® LE 1M and 2M PHY. Support for Volume Control. Support for Scan Delegator to offload Scanning for Broadcast source to a Broadcast assistant. Support for BASS to receive Broadcast Code to decrypt encrypted audio. Support for the CSIP for allowing it to be a member of the Coordinated Set. Support for CAP Acceptor Role.	This role is typically implemented by Hearing Aid devices.
HAUC	Hearing Aid Unicast Client Role. It mandates the following functionality: Support for BAP Unicast Client Role for sending two or more Unicast Audio Streams. Support for CSIP Coordinator Role to discover and pair hearing aids which are binaural and are part of a Coordinated Set. Support for CAP Initiator Role.	This role is typically implemented in devices which are interested in sending Unicast audio to hearing aids, for example a phone, PC, TV, etc.

(Continued)

TABLE 4.2 (*Continued*)
HAP Roles and Functionality

Role	Description of Functionality	Examples
HARC	Hearing Aid Remote Control Role. It mandates the following: VCP Volume Controller Role. Support for CSIP Coordinator Role to discover and pair hearing aids which are binaural and are part of a Coordinated Set. Support for CAP Commander Role.	This role is typically implemented in devices which aid in configuration of the hearing aid or help the user discover nearby Broadcast and select it on behalf of hearing aid devices. Devices like Hearing aid controllers or phones implement this Role.

TELEPHONY AND MEDIA AUDIO PROFILE

The Telephony and Media Audio Profile (TMAP) is designed to aid in the development of interoperable Unicast and Broadcast streaming user experiences for media playback and conversational use cases. It specifies configurations and settings of parameters and procedures that are defined in a lower-level Specification. It does not define new procedures, parameters, or protocols.

Table 4.3 describes the TMAP roles and functionalities.

TABLE 4.3
TMAP Roles and Functionality

Role	Description of Functionality	Examples
CG	The Call Gateway (CG) role is defined for telephony or VoIP applications. The CG device has a connection to the call network infrastructure.	Typical devices implementing the CG role include smartphones, laptops, tablets, and PCs.
CT	The Call Terminal (CT) role is defined for microphone(s)/speaker(s) terminal type applications. A call use case can be made up of a CG and one or multiple CTs.	Typical devices implementing the CT role include wireless headsets, speakers, and microphones that participate in conversational audio.
UMS	The Unicast Media Sender (UMS) role is defined for devices that send media audio content to a sink device in a Unicast Audio Stream.	Typical devices implementing the UMS role include smartphones, media players, TVs, laptops, tablets, and PCs.
UMR	The Unicast Media Receiver (UMR) role is defined for devices that receive media audio content from a source device in a Unicast Audio Stream.	Typical devices implementing the UMR role include headphones, earbuds, and wireless speakers.
BMS	The Broadcast Media Sender (BMS) role is defined for devices that send media audio content to an unlimited number of receiving devices. Enabling Audio sharing use cases.	Typical devices implementing the BMS role include smartphones, media players, TVs, laptops, tablets, and PCs.

(*Continued*)

TABLE 4.3 (*Continued*)
TMAP Roles and Functionality

Role	Description of Functionality	Examples
BMR	The Broadcast Media Receiver (BMR) role is defined for devices that receive media audio content from a source device in a Broadcast Audio Stream.	Typical devices implementing the BMR role include headphones, earbuds, and speakers. A smartphone may also support this role to receive Broadcast Audio Streams from a BMS.

Gaming Audio Profile

The Gaming Audio Profile (GMAP) is designed for low-latency gaming applications over either Unicast or Broadcast transports. It specifies configurations and settings of parameters and procedures that are defined in a lower-level Specification. It does not define new procedures, parameters, or protocols. The profile mandates low-latency settings for both media playback and voice capture, to enable a gaming user experience. GMAP configures these low-latency settings via the Control layer (see Chapter 5).

Table 4.4 describes the GMAP roles and functionalities.

TABLE 4.4
GMAP Roles and Functionality

Role	Description of Functionality	Examples
UGG	The Unicast Game Gateway (UGG) role is defined for gaming and low-latency applications. The UGG device has a connection to either game gateway, call network, or media server.	Typical devices implementing the UGG role include smartphones, laptops, tablets, PCs or game consoles.
UGT	The Unicast Game Terminal (UGT) role is defined for microphone(s)/speaker(s) terminal type applications. A game use case can be made up of a UGG and one or multiple UGTs.	Typical devices implementing the UGT role include wireless headsets, speakers, and microphones that participate in game audio.
BGS	The Broadcast Game Sender (BGS) role is defined for devices that send game media audio content to an unlimited number of receiving devices. Enabling game sharing use cases.	Typical devices implementing the BGS role include smartphones, media players, TVs, laptops, tablets, PCs and game consoles.
BGR	The Broadcast Game Receiver (BGR) role is defined for devices that receive game audio content from a source device in a Broadcast Audio Stream.	Typical devices implementing the BGR role include headphones, earbuds, and speakers.

Public Broadcast Profile

The Public Broadcast Profile (PBP) defines the information metadata of a Broadcast source, such as TV, or public music/media/voice playback, such as a lecture. This profile sets the foundation for Auracast™ broadcast audio, as it defines a Specification that enables the transformation of any Audio Source to a Bluetooth® Broadcaster in a public or personal setting.

The profile specifies the mandatory and recommended streaming parameters that are required for broadcasting the Audio Stream.

This profile is used in a public space, so the metadata and stream are usually not encrypted. The profile may optionally add encryption when the content is considered private to a group of people.

Table 4.5 describes the profile definitions, roles, and their functionality:

TABLE 4.5
PBP Roles and Functionality

Role	Description of Functionality
PBS	The Public Broadcast Source (PBS) extends BAP Broadcast Source which adds ProgramInfo (Program Information) and context type encoding capability. ProgramInfo metadata is one of LTV metadata in Basic Audio Announcement defined at BAP BASE.
PBK	The Public Broadcast sink (PBK) extends BAP Broadcast Sink which adds ProgramInfo and context type decoding capability.
PBA	The Public Broadcast Assistant (PBA) extends BAP Broadcast Assistant which adds ProgramInfo decoding capability.

The profile further defines the program info metadata to provide information about the broadcasted audio, which is part of the Basic Audio Announcement. The profile also advertises the type of Broadcast stream, whether it is high quality (48 kHz) or standard accessible quality (24 kHz), and whether it is encrypted. Table 4.6 describes the metadata information.

TABLE 4.6 Program Info

Metadata	Description of Functionality
Broadcast Name	This is a human-readable string which defines the name of the broadcast.
Audio Active State	It is used to inform the Broadcast sink/assistant whether there is audio present in the Broadcast Audio Stream or not.
Immediate rendering flag	The Broadcast source sets this flag to inform the Broadcast sink to immediately render audio instead of using presentation delay.

REFERENCES

1. HAP version 1.0.1 or later, https://www.bluetooth.com/specifications/specs/hearing-access-profile-1-0-1/
2. TMAP version 1.0 or later, https://www.bluetooth.com/specifications/specs/telephony-and-media-audio-profile-1-0/
3. GMAP version 1.0 or later, https://www.bluetooth.com/specifications/specs/gaming-audio-profile-1-0/
4. PBP version 1.0.1 or later, https://www.bluetooth.com/specifications/specs/public-broadcast-profile1-0-1/

5 Control Layer

In this chapter, we will study the details of the LE Audio Control layer, which is part of the Host protocol stack. We will learn how it enables end-to-end use cases, and configures the parameters used to control the Codec and Transport layers.

ARCHITECTURE

Figure 5.1 illustrates how the Control layer fits in the LE Audio One Architecture. The following subcomponents build the foundation of the Control layer and will be described in detail in the following sections:

1. Audio Stream discovery and establishment
   a. Unicast
   b. Broadcast
2. Volume Control
3. Microphone Control
4. Media Control
5. Call Control
6. Coordination of devices
7. Common Audio procedures

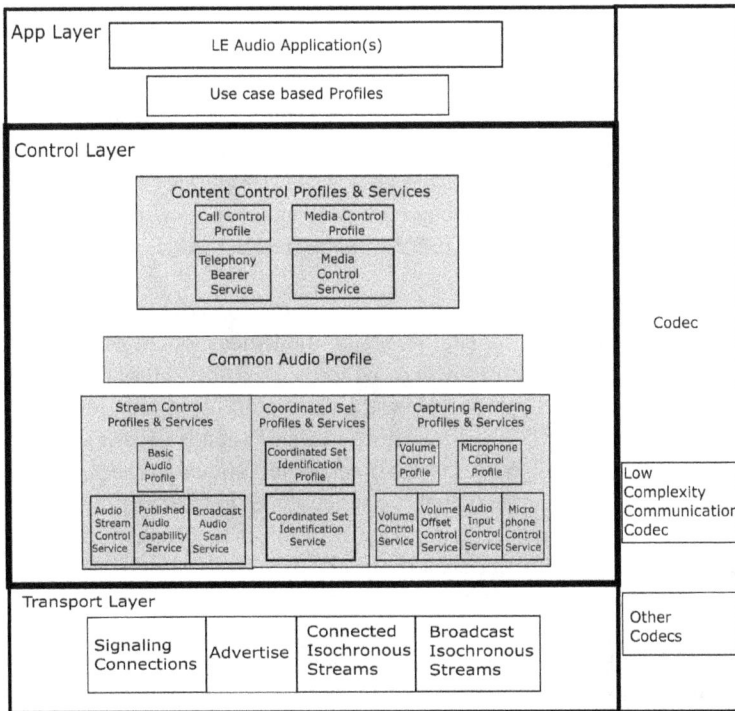

FIGURE 5.1 Architecture of LE Audio Control layer [highlighted boxes].

DOI: 10.1201/9781003590187-5

AUDIO STREAM DISCOVERY AND ESTABLISHMENT

The Audio Stream discovery and establishment is the basic building block of LE Audio. Without this block, none of the audio-related use cases are possible. An Audio Stream can be either Unicast (1:1) or Broadcast (1:many). An Audio Stream represents the manifestation of audio between devices.

UNICAST

This section describes the components, flow, and procedures associated with Unicast Audio Stream discovery and establishment.

Roles

Two distinct roles of Unicast client and Unicast server are defined to separate functionality associated with Unicast Audio Stream establishment. This also enables the discovery, identification, and configuration of devices. Typical examples of Unicast clients include devices like phones, laptops, tablets, and so on. Typical examples of Unicast servers include devices like earphones, speakers, and so on.

Device Discovery

It is the responsibility of the Unicast client to discover the unconnected Unicast server device. This is done using extended advertisements. The Unicast server is a GATT peripheral and advertises ASCS UUID (universal unique identifier), or optionally PACS UUID in the service UUID AD data type of extended Advertising PDU. These UUIDs are defined in Bluetooth® Assigned Numbers. The Unicast server should start with an Advertising interval value of 20 ms to 30 ms for quicker connection setup. If, within 30 seconds, it does not receive a connection from the Unicast client, then it should use an Advertising interval value of 150 ms for power saving.

The Unicast client is GATT central and scans for advertisements from the Unicast server. For quicker connection setup, the Unicast client should use a scan interval of 30 ms to 60 ms and a scan window of 30 ms. If, within the first 30 seconds, it cannot find a Unicast server, then it should use a scan interval of 1.28 s and a scan window of 11.25 ms to reduce power. When it finds a Unicast server in the vicinity, it informs the user and displays the Unicast server device in the user interface for the user to establish a connection. Alternatively, a Unicast client implementation may choose to automatically connect to a Unicast server when it advertises the ASCS UUID, in case it is already bonded with that server. Figure 5.2 illustrates the device discovery procedures between a Unicast client and a Unicast server.

Unicast General and Targeted Announcements may be used by the Unicast server to inform the unconnected Unicast client regarding the availability and imminence of a use case. A Targeted Announcement means that the Unicast server is

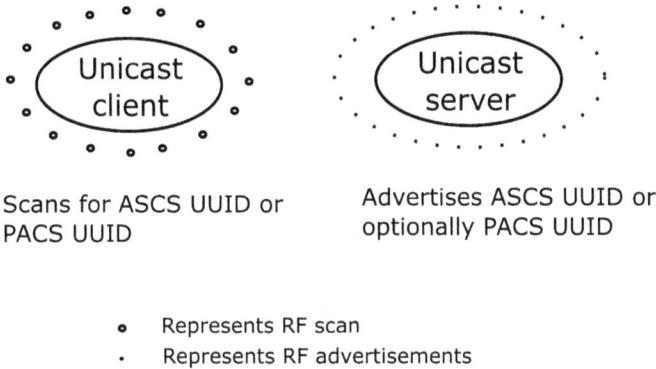

FIGURE 5.2 Unicast client discovering Unicast server.

connectable and is requesting a connection, while a General Announcement means that the Unicast server is connectable but is not requesting a connection. These Announcements are basically extended advertisements with special Bluetooth SIG-defined AD types and availability flag values, which are included in the ASCS UUID service data.

Connection Establishment

When the user selects the Unicast server to establish a connection, then the Unicast client uses any of the following GAP connection establishment procedures based on its preference:

- Auto Connection Establishment procedure: This is used to automatically connect to the Unicast server in the Filter Accept List.
- General Connection Establishment procedure: This allows connection to the specific Unicast server without using the Filter Accept List.
- Selective Connection Establishment procedure: This allows connection to the Unicast server while scanning for multiple devices, selecting a device, and then connecting.
- Direct Connection Establishment procedure: This is used by the server to advertise to a single bonded client and solicit that client to initiate a connection (e.g., via Targeted Announcements).

The Unicast client should use a minimum ACL connection interval of 7.5 ms or 10 ms and a maximum connection interval of 30 ms for a low-latency connection. After that, the connection interval should switch to 50 to 70 ms or to the preferred connection parameters as decided by the Unicast server. Using a low-latency connection allows the profiles to exchange configuration data faster. Increasing the interval allows saving power in the standby connection when profile data is not exchanged. The Bluetooth® SIG added a new Core Controller feature called ECU (Enhanced

Connection Update). The ECU feature allows quicker changes in ACL connection interval to accommodate transitions between low-power standby to low-latency profile signaling, when a user needs to begin/end/update a use case. This feature enables faster responsiveness between devices when switching between low-power standby operation and active use case operation. Chapters 7 and 8 contain more details on this and other new imminent features from the Bluetooth SIG, which will further improve the audio user experience.

Audio Capability Discovery

After successful connection establishment, the Unicast client discovers the audio capabilities of the Unicast server before going ahead with the establishment of the Audio Stream. The Unicast client uses GATT client-based BAP procedures to perform audio capability discovery of the Unicast server. The Unicast server uses a GATT server-based PACS service to expose the audio capabilities. These capabilities are exposed as characteristics of the PACS service, and the Unicast client performs a GATT read of these characteristics.

Some of these capabilities are device-specific, and some are use case-specific. Device-specific capabilities include Audio Location and Audio Contexts, which apply to all the use cases for this device. Use case-specific capabilities include Codec and its related capabilities, which may vary based on the use case. Table 5.1 shows the audio capabilities of the Unicast server, divided between device and per use case.

TABLE 5.1
Audio Capabilities of the Unicast Server

Device-specific capabilities (single instance per device)	Sink Audio Location	
	Source Audio Location	
	Sink Available Audio Contexts	
	Source Available Audio Contexts	
	Sink Supported Audio Contexts	
	Source Supported Audio Contexts	
Use case-specific capabilities (multiple instances)	Sink PAC(s)	Codec ID
		Codec-Specific Capabilities
		Metadata
	Source PAC(s)	Codec ID
		Codec-Specific Capabilities
		Metadata

Audio Location describes the spatial location of the device in case this device is part of a group of devices. Examples of Audio Location include Left, Right, Center, and so on in a multi-speaker configuration. In most cases, the Audio Location of a device is set during factory setup. In some cases, the device may allow configuration of the Audio Location by an application, and therefore these characteristics can also be written using a GATT-write by a Unicast client.

Another aspect of audio capability is Audio Context. Audio Context describes the use case of audio – for example, a telephone call or a media-related use case. It is important for a Unicast client to understand the use cases that a Unicast server is capable of, so that it does not attempt the establishment of a use case that is not supported by a Unicast server. This is specified in the form of the Supported Audio Contexts characteristic, which the Unicast client can read before going ahead with a specific use case with the Unicast server.

Out of the Supported Audio Contexts, the Unicast server may already be engaged in some use cases with another Unicast client at a certain time. In this case, the Unicast server may not be able to support the same use case with another Unicast client. Available Audio Contexts characteristics are exposed to provide this information. The Unicast client can discover the use cases that the Unicast server can support at any specific time by reading the Available Audio Contexts characteristics. In case the Unicast server is already engaged in a use case with some other Unicast client, then this Unicast client can register for notifications for the characteristics. The Unicast server will inform this Unicast client in case there are any changes in Available Audio Contexts using GATT notification. And in case the Unicast server can get engaged in the use case that this Unicast client is interested in, then it can go ahead and establish that use case. Figure 5.3 describes this scenario.

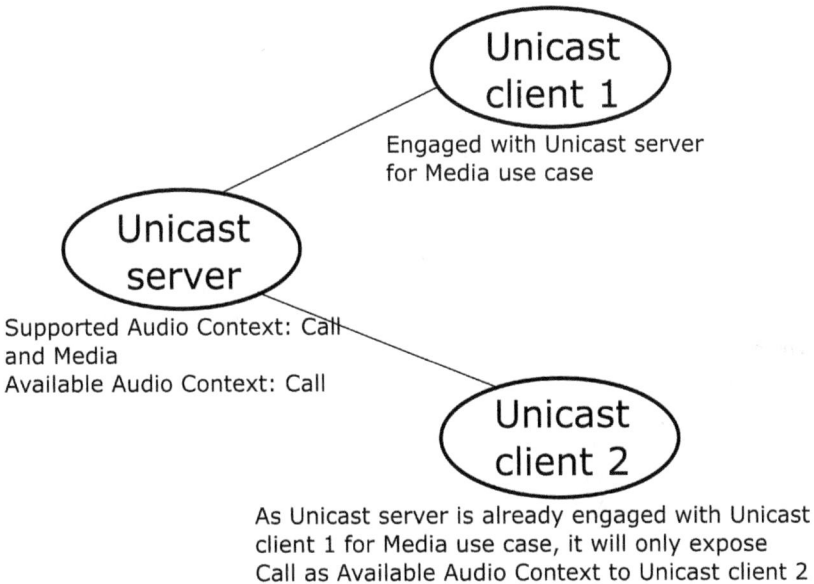

FIGURE 5.3   Illustration of Audio Context capability usage.

The Unicast server exposes the use case-specific audio capabilities using a PAC record. PAC record describes the audio capabilities in a well-established table embedded in the form of a GATT characteristic, which is understood by both the Unicast client and the Unicast server. PAC record primarily exposes the Codec-related capabilities and metadata information. Codec-related capabilities include the Codec ID (Bluetooth® defined or vendor-specific) and its specific capabilities like sampling frequency and so on.

An essential distinction in audio capabilities is whether the device acts as a Source or a Sink – it is called an Audio Role. A Source is where the audio originates (microphone), and a Sink is where audio is consumed (speaker). The PAC records are exposed as Sink and Source PAC, so that the Unicast client is able to quickly retrieve the capabilities it is interested in.

The same is applicable to Audio Location and Audio Contexts. The Unicast server exposes distinct Source and Sink Audio Locations and Source and Sink Audio Contexts.

The Unicast server may only support one of the Audio Roles, and therefore, the other capability may not be present in the exposed characteristic.

Audio Stream Establishment

The Unicast client is responsible for establishing the Audio Stream with the Unicast server. The establishment procedure follows the client–server model, where the Unicast server exposes the Audio Stream-related information using ASCS. All Audio Stream-related information is encapsulated using the Audio Stream Endpoint – which in short is called ASE. And it is represented using the identifier ASE_ID.

The ASCS services expose two types of characteristics for controlling an ASE. The first one is the ASE characteristic, which is used to expose the state of an individual ASE, and there can be more than one of these characteristics on the server, each representing one ASE. There are two types of ASE characteristics: the Sink ASE characteristic and the Source ASE characteristic. The term ASE characteristic is used to refer to both the Sink ASE characteristic and the Source ASE characteristic, wherever they can be used interchangeably (that is, if the description and/or behavior applies to both types of characteristic); otherwise, the characteristics are mentioned by name. The second one is the ASE control point, which is used to control and operate functionality on any of the ASEs on the server. There is only a single instance of this characteristic. ASE_ID is used to distinguish between ASEs. ASE control point operations may act on multiple ASEs in a single operation.

The number of ASEs exposed by the Unicast server is implementation-specific and depends on the number of concurrent Unicast audio Streams that the Unicast server is willing to establish with a single Unicast client. However, the Unicast server may still not be able to support Audio Stream establishment on all of the exposed ASEs simultaneously (due to dynamic resource constraints), and it will reject the establishment of the Audio Stream for that ASE in that scenario. The Unicast client discovers all the ASE characteristics on the server by performing GATT read operations, such as Read by Type.

The Unicast client does a GATT-based write to the ASE control point characteristic for performing BAP Client-related procedures. The ASE control point is also used to inform the Unicast client of a successful (or failed) procedure using GATT-based notifications.

The state of an ASE is managed using a state machine, and Figure 5.4 describes a simplified and abstracted state machine transition diagram for an ASE. The <Codec Configured> and <QoS Configured> states are merged into a single <Configured> state. And the <Enabling>, <Disabling>, and <Releasing> states are optimized out for simplicity. The self-arrow in the <Streaming> state is a context update of a stream via an update metadata operation without disabling the stream. The self-arrow in the <Configured> state allows reconfiguration of an existing Audio Stream for a new use case.

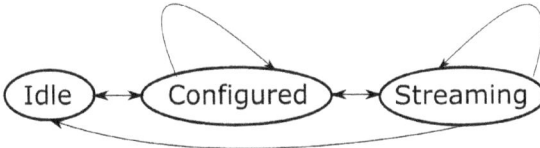

FIGURE 5.4   Simplified and abstracted ASE state machine.

The state machine transitions are primarily caused by the action of the Unicast client. But there are certain cases in which the transitions are also caused autonomously by the Unicast server. The same action or procedure can be applied to multiple ASEs at the same time. The Unicast server maintains a separate state machine per ASE per Unicast client. So, a state machine transition for an ASE ID for one Unicast client does not impact the state machine of the same ASE ID for another Unicast client connected to the same Unicast server.

Idle->Configured State Transition

The <Idle> state signifies that there is no configuration applied to the ASE, and the Audio Stream is not present. The Unicast client configures the Codec and related capabilities by performing the Codec configuration BAP procedure – which leads to the Config Codec ASE control point operation. The information regarding the supported Codec and capabilities is already discovered by the Unicast client during the Audio capability discovery. The Unicast client selects the required value from the set/ range of Audio capability that the Unicast server supports and configures the ASE of the Unicast server with those values. This is based on the use case and directed by the higher layer and the application. The Unicast client also sets initial QoS preference as part of the Config Codec operation; the preferences are sent as Target Latency and Target PHY type. The target latency indicates low, medium, or high latency.

The Config Codec ASE control point operation transitions the state of the ASE to <Codec Configured>, and the ASE control point characteristic is used to notify the Unicast client of this successful transition.

When the ASE is in the <Codec Configured> state, the server exposes the range of QoS parameters that it can support. These QoS parameters reflect the underlying Core configuration that the ASE may be able to support based on the Codec settings and QoS preferences, which were configured in the previous Config Codec step. These QoS parameters include server preference for retransmission number, max transport latency, PHY types, and max SDU size.

The Unicast server also exposes minimum and maximum supported presentation delay values in the <Codec Configured> state. Presentation delay is used for the synchronization of audio across multiple Unicast servers. An example of this use case is left and right earbuds.

The Unicast client selects appropriate values of QoS from the exposed list of QoS parameters and then performs the QoS configuration procedure – which leads to the Config QoS ASE control point operation. The QoS values are based on the use case and selected by the higher layer or application. Along with QoS values, the Unicast client also assigns the CIG and CIS ID of the underlying Core CIS, which is coupled to the Audio Stream of this ASE. Note that these IDs may not have a physical manifestation of the Core CIS yet. The Unicast client configures the retransmission number, PHY type, max transport latency, SDU interval, and max SDU size for this Unicast Audio Stream.

After this procedure, the state of the ASE is transitioned to <QoS Configured>, and the ASE control point characteristic is used to notify the Unicast client of this successful transition. Figure 5.5 shows the sequence of operation for this state transition. After the transition is complete, the server notifies the client about the ASE characteristic state as QoS configured, along with the configured ASE parameters. <Codec Configured> and <QoS Configured> states are abstracted in the <Configured> state in this book, as shown in Figure 5.4.

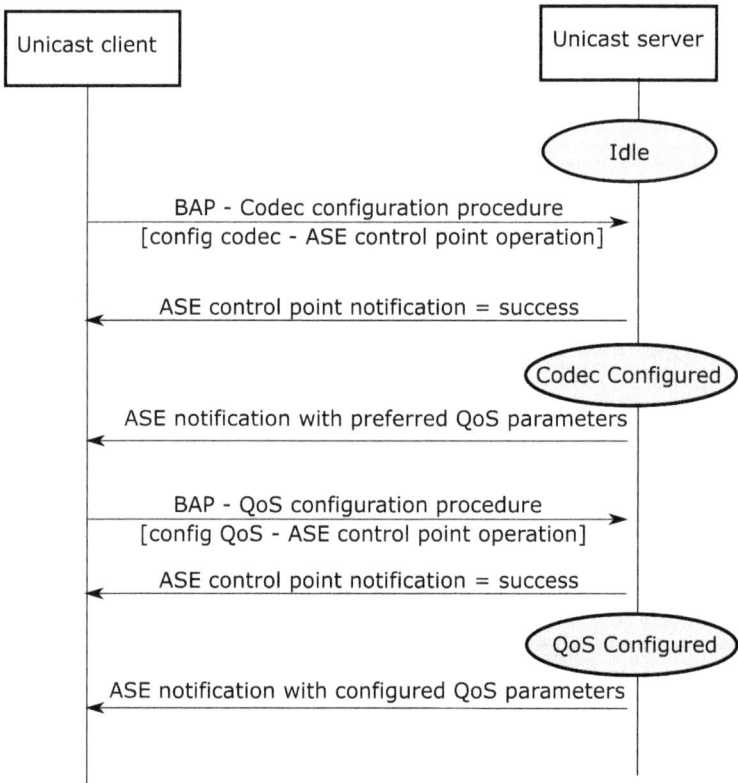

FIGURE 5.5   Idle->Configured state transition sequence.

Configured->Streaming State Transition

The transition to the <Streaming> state involves enabling the ASE and ensuring that the sink side is ready to start receiving audio. Enabling an ASE is a BAP procedure that uses an ASE control point operation by the Unicast client. It also specifies the metadata for the Audio Stream. The metadata may include possible Audio Context that the higher layer or application wants to use this Audio Stream for, and may also include a list of CCIDs. The CCIDs list allows the server to determine which content control service instance it needs to target for this use case for remote control operations such as answering calls, or playing/pausing media. If the operation is successful, an ASE control point notification is sent by the Unicast server to the Unicast client.

The Unicast client then establishes the underlying CIS using the Core Controller and link layer messages. Now the Unicast server transitions to an intermediate state called <Enabling> which indicates the establishment of the CIS, but the CIS may not be carrying actual audio data until the BAP Receiver Start Ready procedure is performed, which is executed by the Unicast client using the ASE control point write operation or by the Unicast server using the ASE state update notification (see next paragraph).

The Receiver Start Ready operation informs a Unicast server acting as the Audio Source (microphone) that the Unicast client is ready to consume audio data transmitted by the Unicast server. The Unicast server, when acting as the Audio Sink (speaker), initiates the Receiver Start Ready operation on its own to inform the Unicast client that the Unicast server is ready to consume audio data transmitted by the Unicast client. This leads to the transition to the <Streaming> state, and actual audio data may be sent over the CIS. In the case of Sink ASE, the server notifies the new <Streaming> state of the ASE to the client, which serves as a Receiver Start Ready notification, while in the case of Source ASE, the client sends an explicit Receiver Start Ready opcode to the server to move the ASE to the streaming state.

Receiver Start Ready is done to ensure that the device that consumes the audio as a Sink has properly initialized its audio-related software and hardware (e.g., audio Codec and DSP) and the buffers are in place to receive audio data from the Audio Source. Else, there are chances that the Audio Sink may miss important audio data from the Audio Source. An example of this is the voice recognition use case, where a phrase (like "Hey Google" or "Hi Siri") is used to invoke a digital assistant. If the Audio Source starts sending audio data without the Audio Sink software and hardware ready to receive this audio data, then the Audio Sink may miss initial parts of the audio data, and the voice recognition may fail. This operation basically denotes that the receiver is ready to start receiving audio data.

Figure 5.6 describes the sequence of operation for this state transition. Once the ASE is in the <Streaming> state, the client may update the stream metadata, for example, when new audio content is deployed over the same stream. This allows using existing LE Audio Streams for multiple use cases without having to remove and recreate streams.

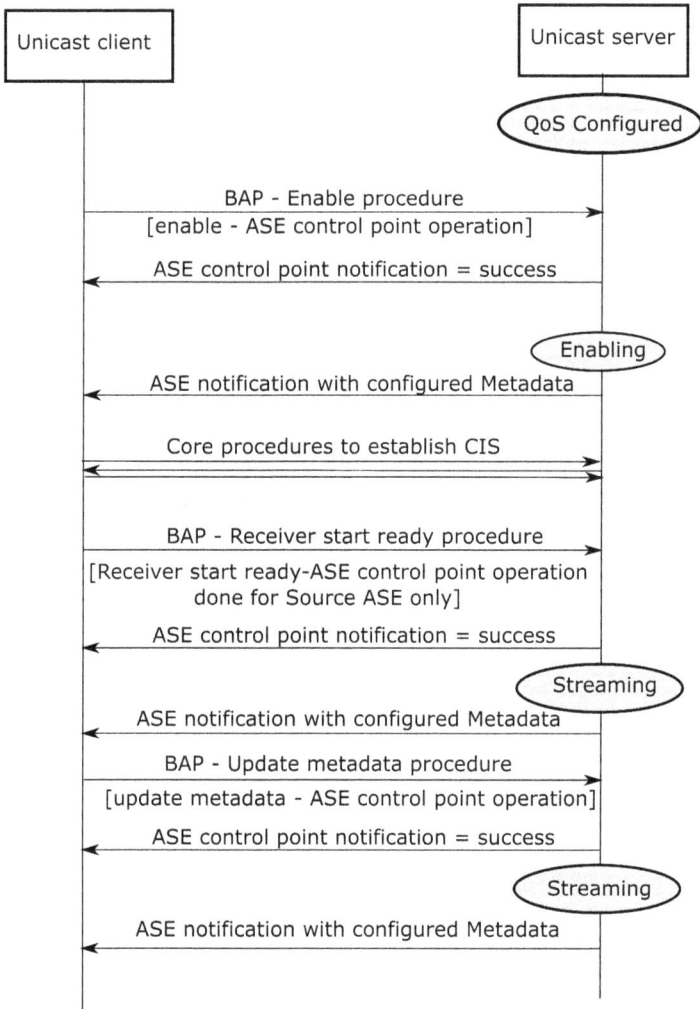

FIGURE 5.6 Configured->Streaming state transition sequence with updating metadata operation during streaming.

Streaming->Configured State Transition

There may be cases where the use case needs to stop streaming audio, yet all of the configuration related to Codec and QoS may still be maintained. This is required for faster enabling of the Unicast Audio Stream and corresponding CIS, when it is required again. The Unicast client may perform the Disable ASE procedure, leading to an ASE control point operation to begin disabling the ASE. Disabling ASE initiates the procedure to stop streaming, but streaming only stops when both sides are ready.

In the case of Source ASE, the Disable ASE procedure moves the ASE to the <Disabling> state. While in the <Disabling> state, the receiver stop ready operation is required to ensure proper de-initialization of the software and hardware of the Audio Sink side (either on the Unicast client or Unicast server) before the audio stops and the underlying CIS is disconnected. As it is possible that the buffering mechanism on the Audio Sink may be looking for audio data from the Audio Source, but CIS termination just pulled the plug without proper synchronization. This operation basically denotes that the receiver is ready to stop receiving audio data.

The receiver stop ready operation is sent from the Unicast client to the Unicast server when it is ready to stop audio from the Source ASE. This operation moves the Source ASE from the <Disabling> state to the <QoS Configured> state. In the case of Sink ASE, the receiver stop ready is done autonomously by the Unicast server by moving directly from the <Streaming> state to the <QoS Configured> state and notifying about the new ASE state. No receiver stop ready operation is done on the Sink ASE. After a successful receiver stop ready operation, when the ASE is moving to the <QoS Configured> state, the Unicast client or the Unicast server may terminate the underlying CIS.

If there are multiple ASEs for the same CIS (denoted by the same CIS_ID and CIG_ID configured in the Enable operation), then the termination of the CIS needs to wait for the receiver stop ready operation to be completed on all of the Source ASEs. An example is Source ASE and Sink ASE, which are mapped to the same bidirectional CIS.

Figure 5.7 describes the sequence of operation for this state transition (Note that the <Disabling> state is skipped for the Sink ASE; no Disabling notification is sent in this case).

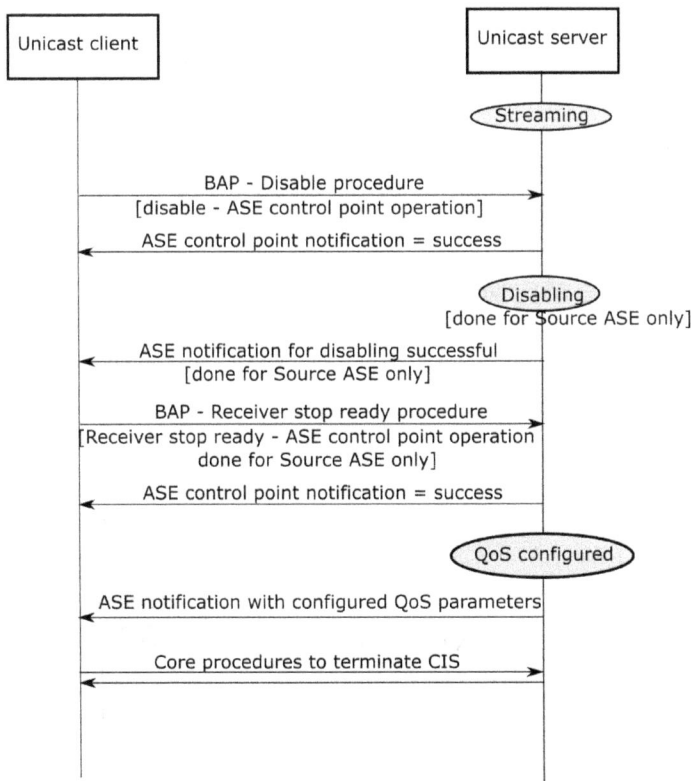

FIGURE 5.7   Streaming->Configured state transition sequence.

Idle State Transition

The ASE may be returned to the <Idle> state in case it is no longer in use. This will enable other applications to use the same ASE for another use case. There is an intermediate state of <Releasing> before the final transition to <Idle>. The <Releasing> state may be reached from any of the other states by performing the BAP Client-initiated Releasing operation and the corresponding ASE control point release operation. It may be performed autonomously by the Unicast server as well.

There may be cases where the Codec configuration of the ASE may have to be cached across connections, and these may have to be persistently associated with the ASE. In this case, the transition from the <Releasing> state is made to the <Codec Configured> state instead of the <Idle> state. It is the prerogative of the Unicast server to either perform a "Released" operation, leading to the <Codec Configured> state of the ASE with all the configuration cached, or transition to the <Idle> state by removing all configurations associated with the ASE. Both the scenarios are shown in Figures 5.8 (with caching) and 5.9 (without caching).

Any mapped CIS is disconnected after releasing an ASE.

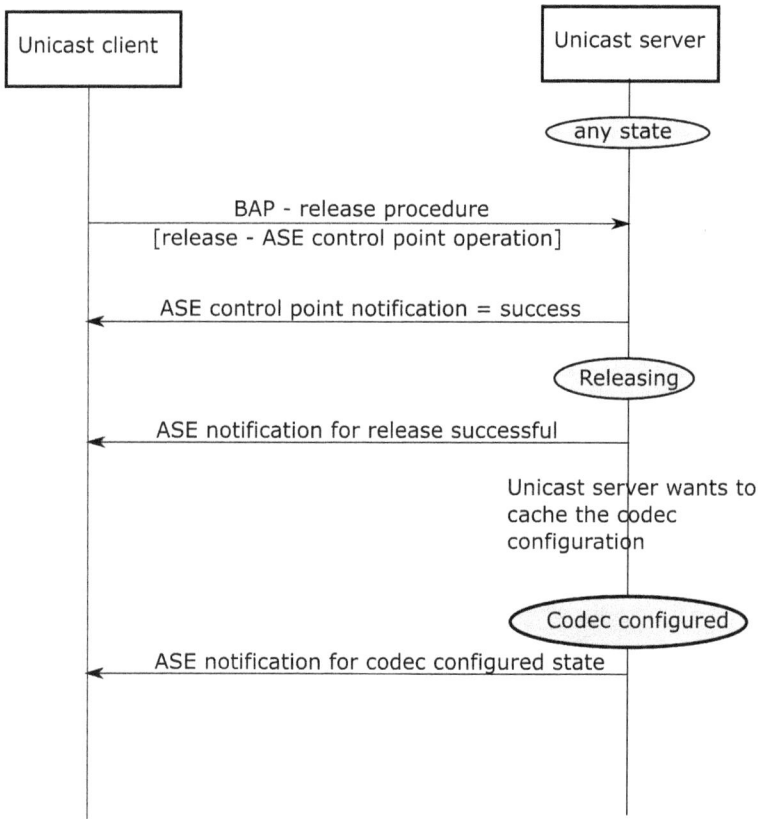

FIGURE 5.8 Release of ASE with caching.

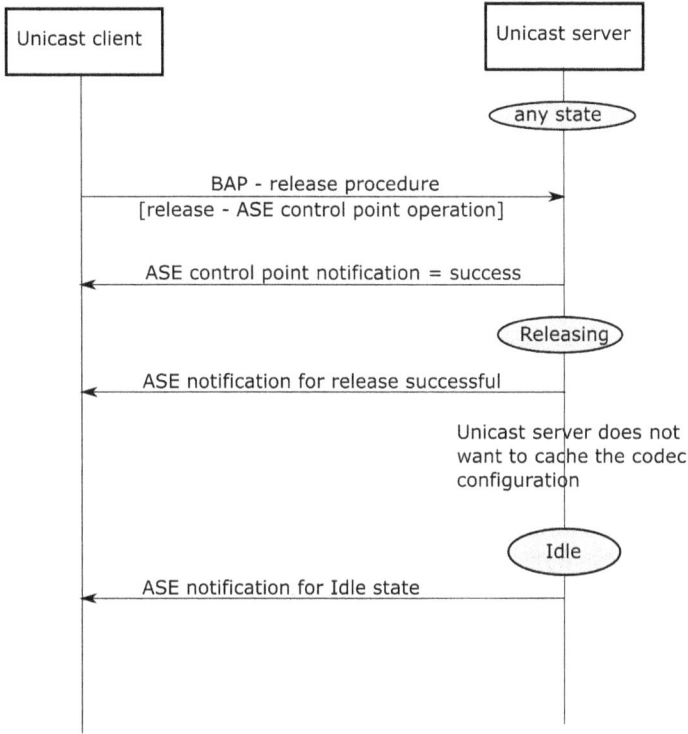

FIGURE 5.9 Release of ASE without caching.

Presentation Delay

For proper rendering of audio across left and right earbuds, both need to be synchronized in time. Presentation delay is a host-level mechanism that is used along with the Core transport latency synchronization mechanism to render synchronized audio across multiple Unicast servers.

Zero Discovery and Zero Configuration

Zero discovery is provided to speed up reconnection. Unicast clients may cache the audio capabilities of the Unicast server and, therefore, do not need to perform the discovery of audio capabilities of the Unicast server after reconnection. This significantly reduces the number of packets exchanged over the air after reconnection and provides a better user experience. It reduces PACS service-related discovery and GATT read of audio capability characteristics.

Unicast clients may also use GATT caching to store the ASE handles of the Unicast server. So the Unicast client may write directly to the cached ASCS control point handle, in order to reconfigure a given cached ASE_ID, knowing the previous ASE state as left after the previous connection. If a given client is leaving the given ASE in the Configured state on a given server, then the client can directly enable the ASE in a later reconnection to the same server. The server keeps separate ASE characteristic states, per remote client, for each ASE_ID.

The combination of client caching, GATT caching, and server ASE state management per client is the concept of zero discovery and zero configuration on reconnection. This enables faster restart of a use case, and streaming can begin with a previous Codec and QoS configuration. An example could be a client that configures one ASE on a given server for music streaming and another pair of ASEs for a voice phone call. At a later time, if the client starts music playback, the music ASE can be enabled immediately. Similarly, if the client wants to make a voice call, the pair of ASEs for the voice call can be enabled immediately, without going through the ASE configuration phase, as all the three ASEs are already preconfigured by the client.

BROADCAST

This section describes the components, flow, and procedures associated with Broadcast Audio Stream discovery and establishment.

Roles

Broadcast functionality is achieved using four distinct roles. Two of the roles – Broadcast source and Broadcast sink – are fairly straightforward to understand and are related to the Broadcast Audio Stream. Broadcast sources are devices that are used for broadcasting information – like a TV in a sports bar, a PC sharing personal audio, an airport announcement kiosk, and so on. Broadcast sinks are devices that are used for receiving broadcasting information, like earphones, speakers, and so on.

Two more roles of Broadcast assistant and Scan delegator are defined to enable resource-constrained devices (like a pair of earbuds) to offload high-power scan functionality to other trusted devices in the vicinity. The other trusted device is called the Broadcast assistant. Typical examples of Broadcast assistants are devices like phones, PCs, or hearing aid remote controls. They are used to scan for Broadcast sources and then pass on relevant information to the Scan delegators, usually implemented along with Broadcast sinks, so that they can receive the Broadcast Audio Stream from a Broadcast source without needing to scan. Figure 5.10 shows how the Broadcast roles are typically implemented in devices.

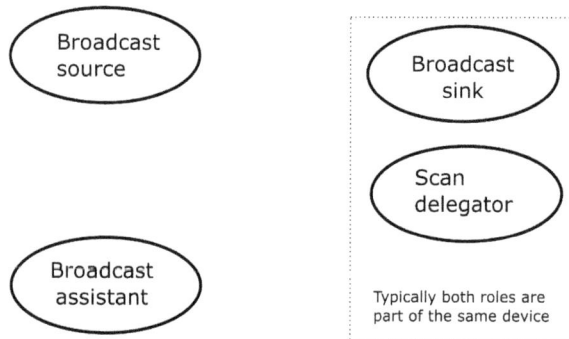

FIGURE 5.10 Broadcast roles.

Device Discovery

The Broadcast source can be discovered by the Broadcast sink or the Broadcast assistant using Broadcast Audio Announcements. These are extended advertisements carrying information about the Broadcast source and the Broadcast Audio Stream. The Broadcast sink or Broadcast assistant scans for extended advertisement, looking for Broadcast Audio Announcement Service UUID (defined in Bluetooth® SIG Assigned Numbers). Even though a Scan delegator may be collocated with the Broadcast sink, it is up to the implementation of the device acting as the Broadcast sink to directly scan or use remote scanning. Figure 5.11 illustrates the discovery of the Broadcast source by either the Broadcast sink or the Broadcast assistant.

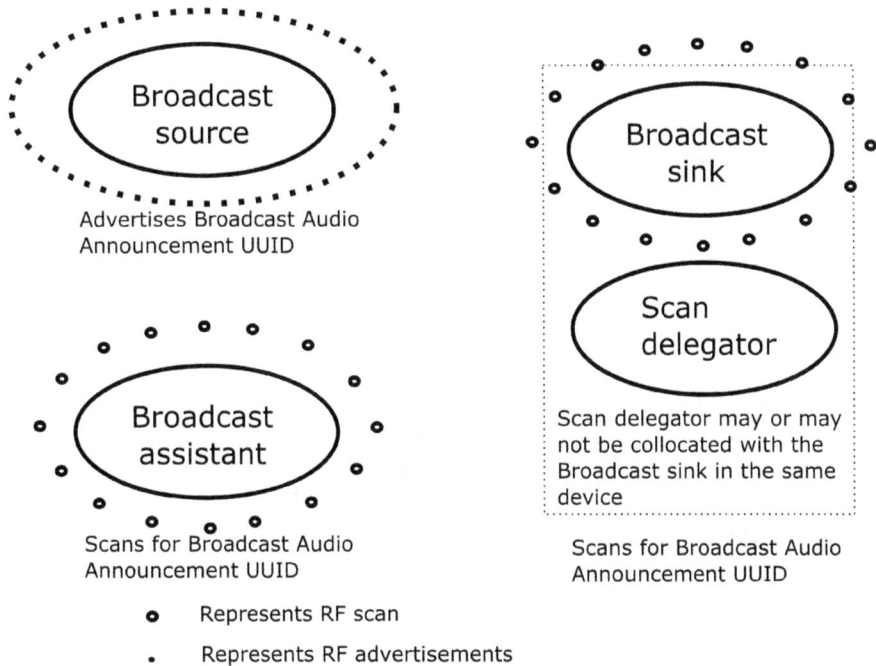

FIGURE 5.11 Broadcast sink or Broadcast assistant discovering Broadcast source.

Scan delegators are discovered by the Broadcast assistant using extended advertisements. If the Scan delegator is interested in offloading the scan to a remote Broadcast assistant device, it starts extended advertisements with the Broadcast Audio Scan Service UUID. The Broadcast assistant, capable of remote scanning, performs scanning of this UUID and discovers the Scan delegator. Figure 5.12 illustrates the discovery of the Scan delegator by the Broadcast assistant.

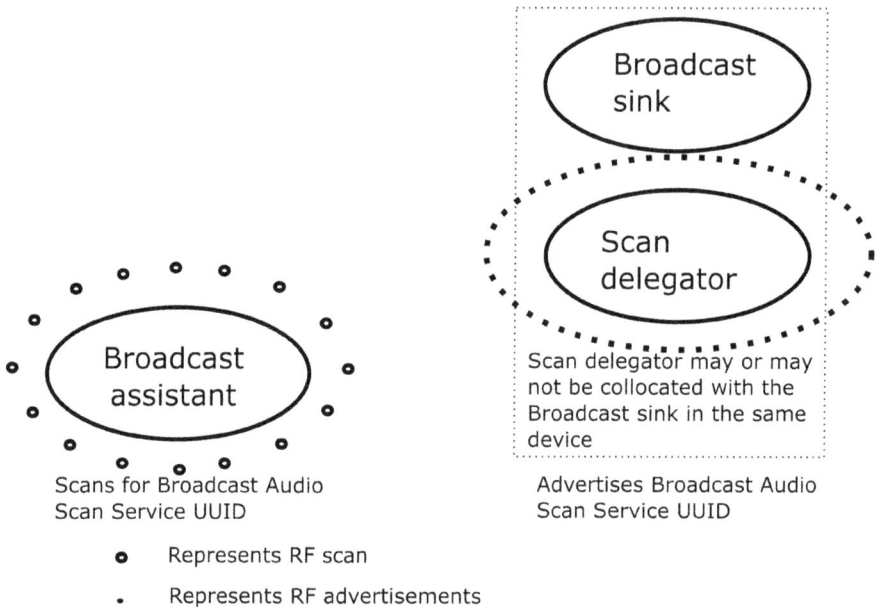

FIGURE 5.12 Broadcast assistant discovering the Scan delegator.

The Advertising interval, scan interval, and scan window mechanism remain the same as in Unicast device discovery.

Connection Establishment

There is no connection established between the Broadcast source and the Broadcast sink. The Broadcast source transmits the Broadcast Audio Stream in a unidirectional connectionless manner, which is received by zero or more Broadcast sinks.

A connection may be established between the Broadcast assistant and the Scan delegator for remote scanning-related procedures. It is the Broadcast assistant that initiates connection establishment with the Scan delegator.

The connection establishment procedures and the minimum and maximum connection interval mechanism remain the same as in Unicast connection establishment.

Audio Capability Discovery

The Broadcast source cannot discover the audio capability of the Broadcast sink and actually does not need to consider the capability of the Broadcast sinks because there may be too many Broadcast sink-based devices in the vicinity, and the Broadcast source will not be able to configure the Broadcast Audio Stream based on each one's capability.

The Broadcast assistant discovers the audio capability of the Broadcast sink. Similar to the Unicast server, the Broadcast sink uses a GATT server-based PACS service to expose the audio capabilities. And the Broadcast assistant uses GATT client-based BAP procedures to discover the audio capabilities of the Broadcast sink. The audio capability discovery procedures remain the same as in the Unicast case, the only difference being that the Broadcast sink can only expose Sink-related PAC and Audio Locations. It also exposes Supported Audio Contexts and Available Audio Contexts in the same manner as in the Unicast case. This enables the Broadcast assistant to select the Broadcast Audio Stream that matches the Audio Contexts that can be currently supported by the Broadcast sink.

Audio Stream Establishment

The Broadcast Audio Stream is established autonomously by the Broadcast source in a unidirectional connectionless manner. All Audio Stream-related information is encapsulated using the Broadcast Audio Source Endpoint – which in short is called BASE. Broadcast Audio Stream establishment is done via Periodic Advertisements and manifested using the underlying BIS (as explained in Chapter 7 in more detail). Periodic Advertising is how the BASE is communicated from the Broadcast source device to the multiple Broadcast sink devices. All Broadcast sink devices will have to sync to the PA from the Broadcast source and read the BASE in order to establish the stream and audio path to their speakers. When BIG/BIS is set up by the Broadcast source, the PA also contains a BIGInfo element to direct the Broadcast sink devices to the timing and scheduling parameters of the BISes PDUs.

The properties of the BIS and the Broadcast Audio Stream are exposed using Basic Audio Announcements and the BASE structure. Basic Audio Announcements are Periodic Advertisements characterized by the Basic Audio Announcement UUID.

Broadcast Audio Announcements in the extended advertisements have a SyncInfo field. The presence of the SyncInfo field in an extended advertisement indicates the presence of a PA train. The contents of the SyncInfo field describe the timing and info about the PA train. And this PA train contains the Host Basic Audio Announcement with the BASE structure. Figure 5.13 illustrates the link between the Broadcast Audio Announcement (contains UUID) and the Basic Audio Announcement (contains UUID and BASE). The BASE in PA is what is required in order to synchronize to the BIS Broadcast stream; therefore, Broadcast sinks that offload scan only need to synchronize to the PA train with BASE info and BIGInfo.

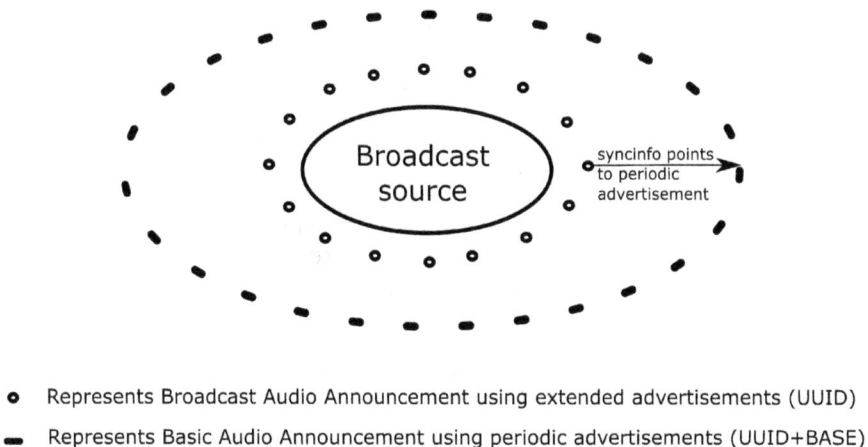

o   Represents Broadcast Audio Announcement using extended advertisements (UUID)

▬   Represents Basic Audio Announcement using periodic advertisements (UUID+BASE)

FIGURE 5.13   Link between Broadcast and Basic Audio Announcements.

BASE is a hierarchical structure loosely based on the PAC record and organized in an object-oriented style, borrowing concepts of inheritance and polymorphism. Using the concept of inheritance, the base structure of BIG defines the common properties of all the BISes that are part of the BIG. Each BIS derived from this BIG has the same properties unless overridden using the polymorphism concept. The BISes can also be organized into subgroups to allow for grouping of BISes with the same properties. This hierarchy is used to save over-the-air space for the PA PDU. The same Periodic Advertisement also carries the link layer control and timing information for the pointed BIG. Figure 5.14 shows the BASE hierarchical organization.

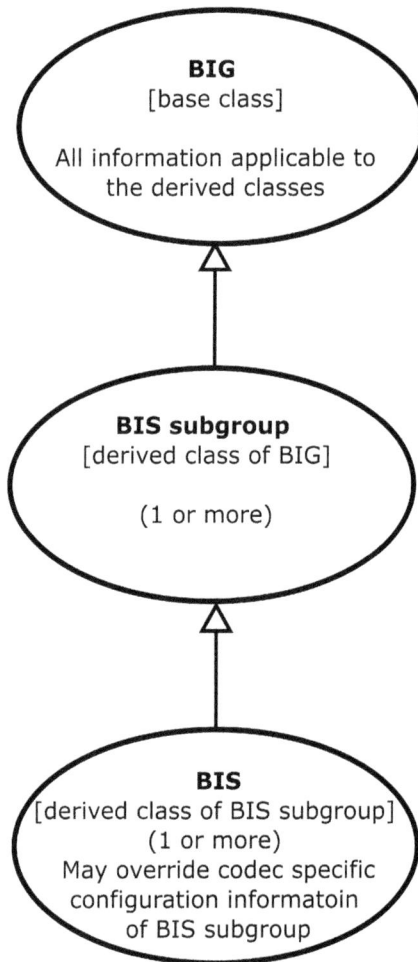

FIGURE 5.14 BASE explained in object-oriented terminology.

BIGInfo is the difference between whether the Broadcast Audio Stream is in the <Configured> state or in the <Streaming> state. If BIGInfo is not present in the ACAD (Additional Controller Advertising Data) field of the PA train, then the Broadcast Audio Stream is in the <Configured> state, and there is no actual BIS that is set up and no Broadcast Audio is transmitted by the Broadcast source. If a BIGInfo field is present, then the BIS is set up and the Broadcast Audio Stream is in the <Streaming> state. And the actual audio data is transmitted by the Broadcast source.

The Broadcast sink or the Broadcast assistant can scan for Basic Audio Announcements and then start receiving the PA report with Host Advertising data, which contains the BASE structure. The BASE structure describes the audio configuration. BIGInfo reports are also sent to the host profile. The BIGInfo reports are an additional part of the PA, as part of ACAD. When a Broadcast sink needs to sync to a BIG, it uses the information it received in the BASE and the BIGInfo to guide its controller to sync to the BIG stream. The BIGInfo contains synchronization information about the various BIS indices in the BIG, and the BASE describes the audio configuration of each BIS index. A Broadcast sink may sync to a single BIS from the BIG or to more than one BIS, depending on each BIS content and the sink audio capabilities from the PAC records.

Remote Scanning

Remote scanning is scanning for the Broadcast sink by the Broadcast assistant on behalf of the Scan delegator. This is a feature-rich set of procedures where the Broadcast assistant can discover and operate on two characteristics exposed by the GATT-based Broadcast Audio Scan Service (BASS) of the Scan delegator. One of the characteristics is to provide the state of the Broadcast sink, which is called the Broadcast receive state. This characteristic is read by the Broadcast assistant and is also notified to it, if there are changes. The second is the Broadcast Audio Scan Control Point, which is primarily used to operate on the Broadcast receive state characteristic to execute a set of remote scanning procedures. There can be only one Broadcast Audio Scan Control Point, while there can be more than one Broadcast receive state characteristics exposed by the BASS.

These are the primary functionalities provided by the Broadcast assistant:

- The Broadcast assistant can start scan or stop scan on behalf of the Scan delegator.
- Based on user preference, the Broadcast assistant can request the Scan delegator to add, modify, or remove a Broadcast source, which the Broadcast assistant has scanned for. When it does request to add a Broadcast source, it may lead to the PAST, which is described in the next section.

- Some Broadcast Audio Streams may be encrypted, and the Broadcast assistant can also provide encryption key information (called the Broadcast code) to the Scan delegator for decrypting the Broadcast Audio Stream.

The BAP Client and the BASS service define a set of client–server operations and procedures to carry out the preceding functionality.

Scan Offloading Using PAST

Scan offloading essentially means the transfer of SyncInfo information from the Broadcast assistant to the Scan delegator using the PAST procedure. After the Broadcast assistant has performed remote scanning on behalf of the Scan delegator, it can actually request the Scan delegator to begin a sync to the Broadcast source that the Broadcast assistant scanned for and the user has selected. The PAST is a controller link layer procedure, which is initiated by the Host. Once the Host initiates the procedure, the Controller exchanges the PA timing information on how to sync to the PA with the Basic Broadcast Announcement (BASE) and BIGInfo of interest.

The Broadcast assistant invokes the add Broadcast source procedure, and on successful completion of this procedure, the link layer sends the LL_PERIODIC_SYNC_IND with the SyncInfo and the PA Sync Offset to the Scan delegator. The Scan delegator can then use this SyncInfo and Sync Offset to sync to the PA train and then directly receive the Periodic Advertisement reports containing the BASE and Broadcast Audio Stream information and thus receive the Broadcast audio timing (BIGInfo) from the Broadcast source. The Scan delegator can later use the BIGInfo timing info to sync to the BIG from the Broadcast Audio Source and start streaming audio with Codec settings as described in the BASE.

VOLUME CONTROL

Volume control is used for controlling the state and volume of the speaker associated with the device. It is implemented using the GATT client-based Volume Control Profile (VCP) and GATT server-based Volume Control Service (VCS), and an optional Volume Offset Control Service (VOCS) and optional Audio Input Control Service (AICS). VCS is required to control the relative as well as absolute volume of the device. VOCS is required to control the volume offset between speakers present in the same device to balance and equalize multiple speakers in a device. AICS is required to control other inputs to the device speaker, such as HDMI, USB, or an internal volume source from a microphone. VCS is the primary service, and VOCS and AICS are secondary services (if implemented).

The VCS, VOCS, and AICS services are implemented in devices that have the actual speaker for rendering audio, like earphones, speakers, and so on. The VCP profile is implemented in devices (like phones, laptops), which are typically used to control the volume of these peripherals. This separation of functionality is done in terms of well-defined roles – Volume Controller and Volume Renderer. Figure 5.15 shows the relationship between the roles, the profiles, and the services.

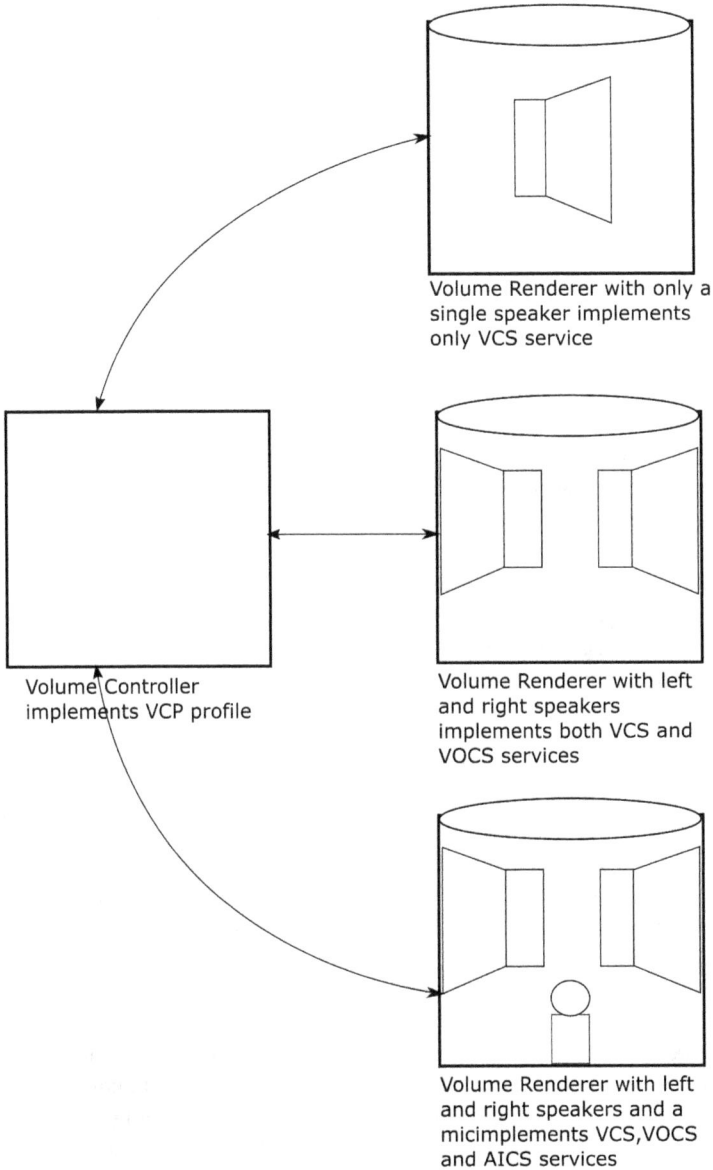

FIGURE 5.15 Illustration of volume control roles, profile, and services relationship.

The procedures and operations are executed by the Volume Controller using a GATT-based control point.

The following functionality is supported by the volume control block in VCS:

- Setting up the initial volume of the Volume renderer
- Configuring mute and unmute of the Volume renderer
- Configuring relative volume up and down without changing the mute state

- Configuring relative volume up and down with a change to the mute state, that is, volume up or down will also unmute the Volume renderer, if it is muted
- Setting the absolute volume of the Volume renderer

The following functionality is supported by the volume control block in VOCS:

- Read and write the Audio Location and Audio Description of the independent speaker in the device
- Configuring the volume offset of the independent speaker in the device required for balancing and equalizing

The following functionality is supported by the volume control block in AICS:

- Read and write the Audio Description of the independent volume input sources in the device. Examples of Audio Descriptions are "Bluetooth®" or "Line-in." This is to identify the type of independent input volume source attached to the device speaker, such as a built-in microphone or an external line-in wire.
- Configure gain mode – either manual or automatic. In the case of automatic gain mode, the other input device or microphone adjusts the gain without the need for manual configuration.
- Configure gain settings of the independent input or microphone in the device, in case it is configured for manual gain mode.
- Configure mute and unmute of the independent input source or microphone in the device.

Additionally, a change counter-based operation is provided so that the Volume Controller and Volume Renderer are always in sync. For each operation, the change counter is also included by the Volume Controller. If the change counter in the operation does not match the change counter of the Volume Renderer, then the Volume Renderer returns an error. In that case, the Volume Controller needs to read the latest state and volume settings of the Volume Renderer and retry the operation with the latest change counter. The purpose of the change counter is to handle cases when more than one Volume Controller client exists. For example, there could be three different remote controls used to make volume changes to a single Bluetooth technology-based speaker. If the user starts using remote control A, and then later switches to use remote control B or C, and later goes back and uses remote control A, then the remote control A must sync to the current volume settings done by remote control B or C before it can begin making volume changes. Otherwise, if remote control A's last setting shows volume 3, while the actual volume level is 7, and if the user wants to raise the volume level to 7 (i.e. an increment of 4) as per remote control A, then the resulting actual volume level will go to 11 on the Bluetooth technology-based speaker, while remote control A shows a volume level of 7. The change counter prevents such cases by detecting a controller that is out of sync and requires it to sync to the current settings by reading the volume states again, including the latest change counter value.

MICROPHONE CONTROL

Microphone control is used for controlling the state and gain of the microphone associated with the device. It is implemented using the GATT client-based MICP and GATT server-based MICS, and an optional AICS. MICS is required if the only feature required is a device-wide mute control. AICS is required in case the device supports advanced microphone-related features, like gain control of the microphone and independent microphone control (for devices like mixers, which can have multiple microphones – either wired or wireless). MICS is the primary service, and AICS is the secondary service (if implemented).

The MICS and AICS services are implemented in devices that have the actual microphone for capturing audio, like headsets, wireless microphones, mixers, and so on. The MICP profile is implemented in devices (like phones, laptops), which are typically used to control the gain of these peripherals. This separation of functionality is done using well-defined roles – Microphone Controller and Microphone Device. Figure 5.16 shows the relationship between the roles, the profiles, and the services.

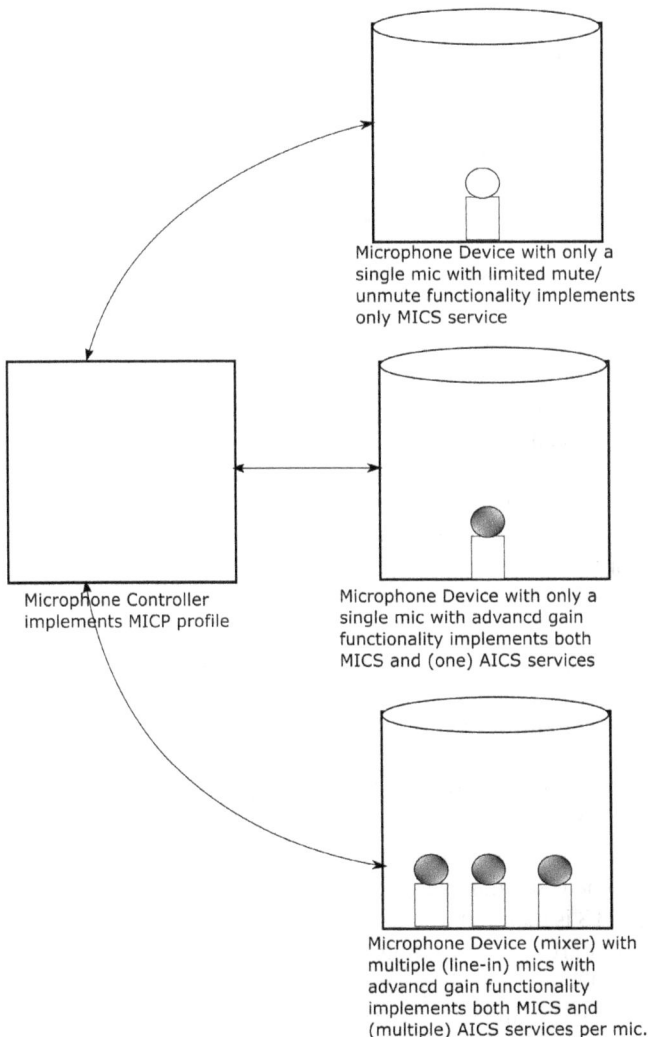

Microphone Device with only a single mic with limited mute/ unmute functionality implements only MICS service

Microphone Controller implements MICP profile

Microphone Device with only a single mic with advancd gain functionality implements both MICS and (one) AICS services

Microphone Device (mixer) with multiple (line-in) mics with advancd gain functionality implements both MICS and (multiple) AICS services per mic.

FIGURE 5.16 Illustration of microphone control roles, profiles, and services relationship.

The procedures and operations are executed by the Microphone Controller using the GATT-based control point.

The following functionality is supported by the microphone control block in MICS:

- Configuring mute and unmute of the Microphone device

The following functionality is supported by the microphone control block in AICS:

- Read and write the Audio Description of the independent microphone in the device. Examples of Audio Descriptions are "Bluetooth®" or "Line-in." This is to identify the type of independent microphone attached to the device.
- Configure gain mode – either manual or automatic. In the case of automatic gain mode, the external line-in or microphone adjusts the gain without the need for manual configuration.
- Configure gain settings of the independent microphone, or external input line-in, in the device, in case it is configured for manual gain mode.
- Configure mute and unmute of the independent, external line-in or built-in microphone in the device.

Similar to Volume control, a change counter-based operation is provided so that the Microphone Controller and Microphone Device are always in sync. For each operation, the change counter is also included by the Microphone Controller. If the change counter in the operation does not match the change counter of the Microphone Device, then the Microphone Device returns an error. In that case, the Microphone Controller needs to read the latest state and microphone settings of the Microphone Device and retry the operation with the latest change counter. The change counter helps to synchronize multiple Microphone Controller clients. For example, if a wireless microphone is controlled by two remote controls and remote control A mutes the microphone, then this change needs to propagate to remote control B, so remote control B can correctly reflect to the user that the microphone is currently muted. The change counter allows remote control B to realize that it must read the latest changes before attempting any controller operation, such as unmute or mute.

MEDIA CONTROL

Media control is used for controlling media player-related functionality (e.g., play, pause, rewind, forward, track changes). It is implemented using the GATT client-based MCP and GATT server-based GMCS, an optional MCS, and an optional Object Transfer Service (OTS). GMCS is required when the device is treated as a single player and generic media player control functionality is required. MCS is required when there are multiple specific players in the device, and each one needs its own control. OTS is required whenever there is object-based information exchange – like album cover art, icon, track segments, and so on. MCS and GMCS are primary services, and OTS is a secondary service (if implemented).

The MCS, GMCS, and OTS services are implemented in devices that have the actual media players playing media/audio, like phones, laptops, and so on. The MCP profile is implemented in devices (like headsets, controllers), which are typically used to remotely control these media players. This separation of functionality is done in terms of well-defined roles – Media Control Client and Media Control Server. Figure 5.17 shows the relationship between the roles and the profiles and services.

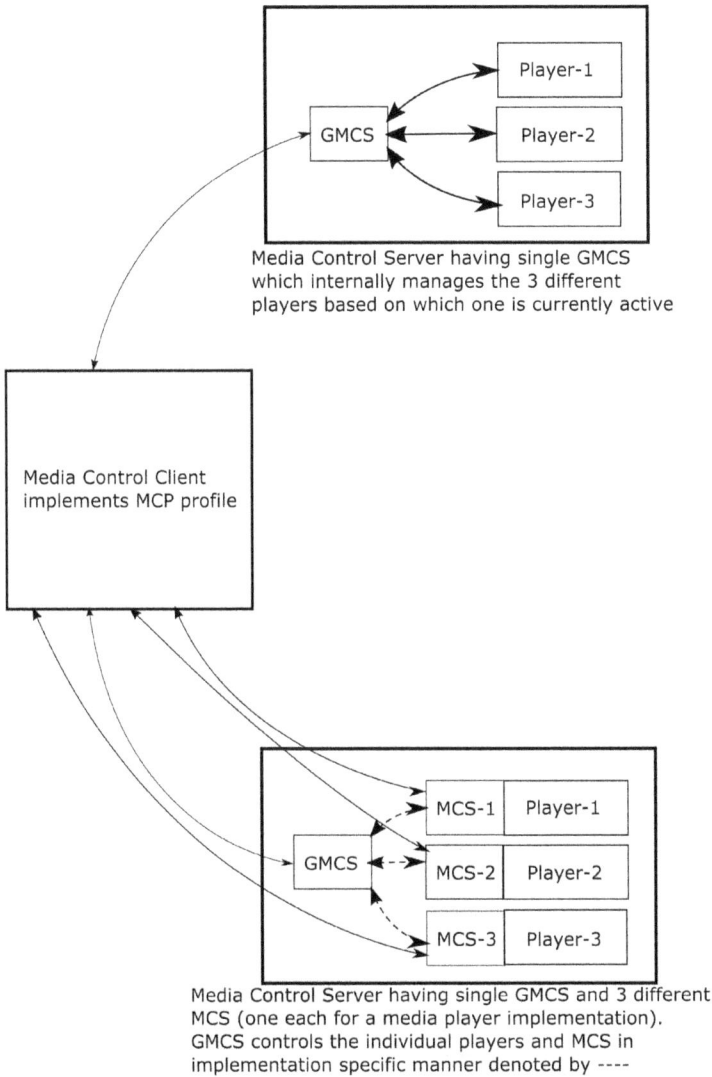

FIGURE 5.17   Illustration of media control roles, profiles, and services relationship.

The procedures and operations are executed by the Media Control Client using the GATT-based control point.

The MCS media information is structured in the form of a hierarchy of groups, tracks, and segments. There can be multiple groups stored at the MCS server, each group having multiple tracks, and each track is subdivided into multiple segments. The media control procedures allow an implementation to navigate the media information using the group-, track-, and segment-related structure. At any point in time, a current segment of a current track of a current group would be playing. However, implementing these is optional and based on the requirements of the media player implementation. Figure 5.18 shows the conceptual organization of groups, tracks, and segments. Examples for groups could be a music album, a podcast name, or a movie title. Examples for tracks could be album singles, podcast chapters, or a movie scene. Examples for segments could be a part of an album single, a section from a podcast chapter, or a section from a movie scene. The organization of groups into tracks and into segments is implementation-specific and content-specific, for example, a track may contain a single segment, and a group may contain a single track.

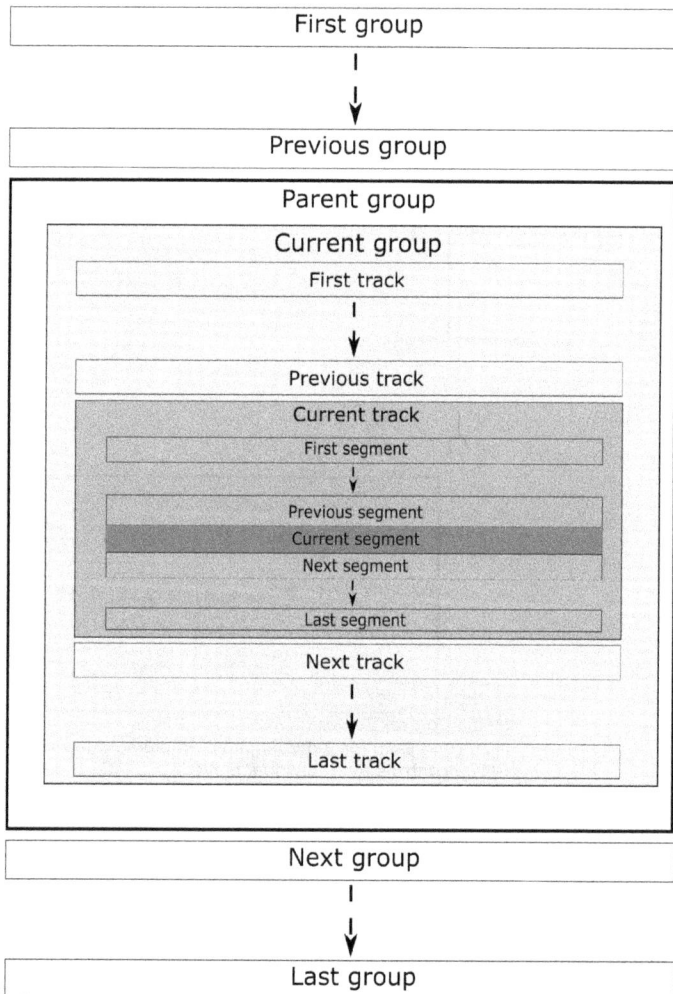

FIGURE 5.18 Illustration of MCS hierarchy groups, tracks, and segments.

The following functionality operations are supported by media control:

- Media Player-related operations:
- Media Player refers to the various types of Media Players that can play the media/music content at the Media Control Server-based device.
  - Read Name: The Media Control Client can read Media Player Name, which is an implementation-specific UTF-8-based text. Examples are some of the popular music players like "Google Play," "Amazon Music," "Spotify," and so on. GMCS should expose the name of the media player that is currently active. If there is any change in the Media Player, then a notification is sent to the Media Control Client. The Media Control Client then reads the information regarding the Media Player again.
  - Read Icon: The Media Control Client can read the Media Player Icon Object, which is an implementation-specific icon of the media player in .bmp format. It uses the OTS service to receive this object.
  - Read Icon URL: The Media Control Client can read an optional Media Player Icon URL and download the icon of the Media Player directly from the Internet.
- Group-related operations:
  - Discover by Group Object
  - Set current Group Object ID
  - Read Parent Group Object information
  - Move to First Group
  - Move to Last Group
  - Move to Next Group
  - Move to Previous Group
  - Move to Group Number
- Track-related operations:
  - Move to First Track
  - Move to Last Track
  - Move to Next Track
  - Move to Previous Track
  - Move to Track Number
- Segment-related operations:
  - Move to First Segment
  - Move to Last Segment
  - Move to Next Segment
  - Move to Previous Segment
  - Move to Segment Number
- Current Track-related operations:
  - Play
  - Pause

- Fast Forward
- Fast Rewind
- Stop
- Read Title
- Read Duration
- Read Position
- Set Absolute Position
- Set Relative Position
- Read Segments Object information
- Read Object ID
- Read Object information
- Next Track-related operations:
  - Read Object information
  - Set Object ID
  - Set Object information
- Playback Speed-related operations:
  - Read
  - Set
  - Read seeking speed
- Playing Order-related operations:
  - Read
  - Set
  - Read Playing Order supported
- Search-related operations (seek)
- Read Content Control ID-related operation

TELEPHONY CONTROL

Telephony control is used for controlling telephony-related functionality (e.g., accept/reject call, redial). It is implemented using the GATT client-based CCP and GATT server-based GTBS, and an optional TBS. GTBS is required when the device is treated as a single telephony device and generic telephony control functionality is required. TBS is required when there are multiple telephony clients (e.g., cellular-based or VoIP-based) in the device, and each one needs its own control. TBS and GTBS are primary services.

The TBS and GTBS services are implemented in devices that have the actual telephony clients and applications, like phones, laptops, and so on. The CCP profile is implemented in devices like headsets and controllers, which are typically used to remotely control telephony clients and applications. This separation of functionality is done in terms of well-defined roles – Call Control Client and Call Control Server. Figure 5.19 shows the relationship between the roles and the profiles and services.

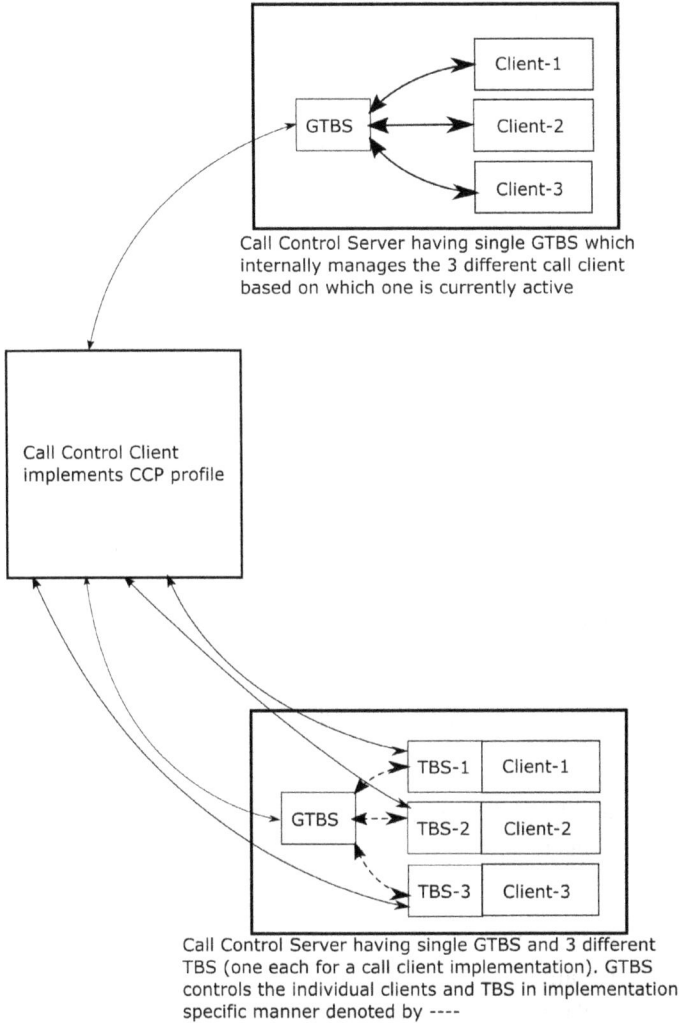

Call Control Server having single GTBS which
internally manages the 3 different call client
based on which one is currently active

Call Control Client
implements CCP profile

Call Control Server having single GTBS and 3 different
TBS (one each for a call client implementation). GTBS
controls the individual clients and TBS in implementation
specific manner denoted by ----

FIGURE 5.19   Illustration of call control roles, profiles, and services relationship.

The procedures and operations are executed by the Call Control Client using the
GATT-based control point.

The following functionality operations are supported by the TBS and GTBS:

- Bearer-related operations
- A Telephony bearer is defined as the actual backbone over which the
  telephony-related information exchange would be performed. It could be
  a GSM-based, 4G or 5G Cellular bearer performing standard circuit or
  packet switch-based network protocols. It could also be other Voice over
  IP (VoIP)-based applications (like Teams, Skype, Zoom, WhatsApp call).

There is a possibility of telephony-related calls active in multiple bearers at the same time, and each of them may have a corresponding TBS service instance. The following are bearer-related procedures and operations:

- Read Provider Name: This is a friendly name of the Bearer.
- Read UCI (Uniform Caller Identifiers): This is Bluetooth® SIG Assigned Numbers-based standard names of the bearers. This is defined in www.bluetooth.com/specifications/assignednumbers/uniform-caller-identifiers/.
- Read Technology: This is the type of technology (e.g., 3G, 4G, LTE, and Wi-Fi) that is being used for this bearer. This is defined in Bluetooth SIG Assigned Numbers.
- Read URI (Uniform Resource Identifier) schemes supported list: These are the URI schemes that this bearer supports. They are defined in www.iana.org/assignments/uri-schemes/uri-schemes.xhtml.
- Read Signal Strength: This is a value between 0 and 100, which indicates the implementation-specific signal strength of the bearer.
- Read Signal Strength Reporting Interval
- Set Signal Strength Reporting Interval
- Read List Current Calls
- Read Incoming Call Target Bearer URI
- Call-related operations:
  - Read Status: The status of the Call Control Server is In-band Ringtone enabled or disabled, and whether the Server is in Silent mode or not.
  - Read Call State: The state of each call in an implementation-specific index-based array. The following are the possible values of the call state: Incoming, Dialing, Alerting, Active, Locally held, Remotely held, and Locally and remotely held.
  - Accept Incoming Call
  - Terminate Call
  - Move Call to Local Hold
  - Move Locally Held Call to Active
  - Move Locally and Remotely Held Call to Remotely Held
  - Originate Call
  - Join Calls
- Read Content Control ID operation

COORDINATION OF DEVICES

Coordination of devices is a mechanism to identify and treat multiple Bluetooth® LE devices as part of a single coordinated set. For LE Audio, it enables ease of discovery and pairing of independent yet synchronized devices, like left and right earbuds or multiple speakers of an N.1 digital home theater system.

The GATT-based CSIP and Service (CSIS) are designed to enable this. To be identified and discovered as part of a coordinated set, the primary requirement is to share a key (called SIRK – Set Identity Resolving Key) among devices that are required to be part of this coordinated set. The mechanism to share and provision this key is out of scope of these Specification. It could be done during manufacturing, or out-of-band provisioning mechanisms may be used. The purpose of the key is to allow generating different identifiers for set members, which are a hash function of the same set shared key. This way, multiple set devices with different addresses are Advertising an identifier that can be resolved into the same set key.

The device that authenticates to the set of devices is called the set coordinator, and it is the CSIP client. The various devices that form the set are called set members and implement a CSIS server.

The SIRK is the set key, and it is used to generate and resolve a RSI, which reduces the ability to track this device because it changes frequently. To discover a coordinated set and its members, the first step is to discover at least one member of the coordinated set. After discovering, connecting, and authenticating to the CSIS server at the set member, the CSIP client device reads the SIRK from the set member. This SIRK is then used to discover other members of the set, by resolving more RSI values, which are advertised by the other members of the coordinated set.

Figure 5.20 summarizes the set members' discovery. The figure shows one set coordinator that authenticates to two set members. An example can be a phone or a PC, which authenticates to a pair of earbuds.

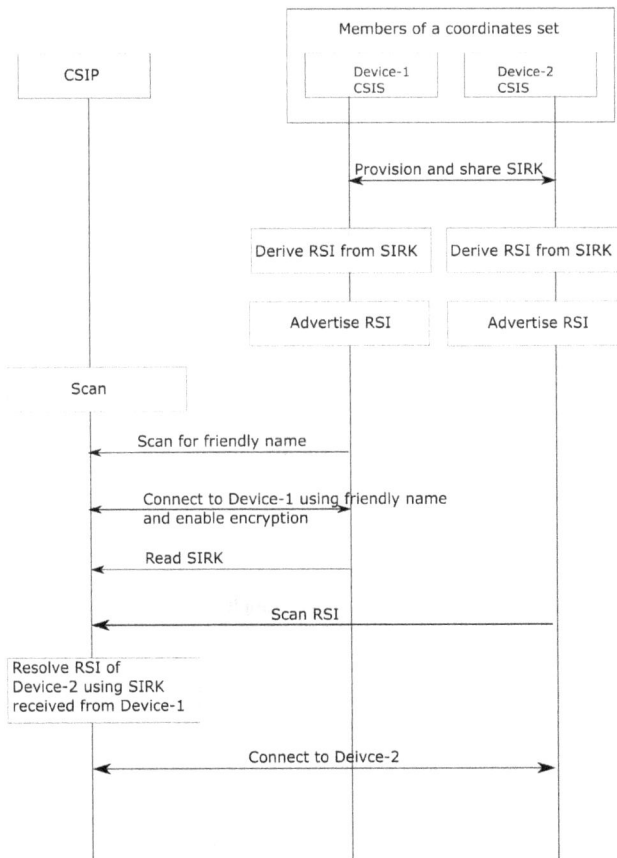

FIGURE 5.20 Discovery and connection of members of the coordinated set.

The SIRK reading and RSI resolution are done in addition to Bluetooth pairing and authentication. The pairing and authentication are 1:1 procedures, with each of the set members. Each connection to a set member is separately authenticated and encrypted. Authenticating each set member is essential to prevent impersonation. For example, a third device may try to apply impersonation and pretend to be a set member and then eavesdrop on a voice call. By mandating authentication of each set member, the third device is rejected from joining the set. Figure 5.20 assumes that each connection is authenticated by each set member. An example of authentication by a set member could be a button press on each earbud or a voice gesture ("Do you agree to connect now to this source?"). The user agrees to connect only when they trust the CSIP client at that moment. When such authentication is applied, the third impersonating device fails to read the SIRK when it tries to pretend to be a set coordinator; and after intending to read the SIRK from an existing valid set member, it may later pretend to be a set member using the same SIRK it read.

COMMON AUDIO PROFILE

The Common Audio Profile (CAP) defines common procedures that are followed by the application profiles. The application profiles use the common audio procedures when applying multiple controls over a set of audio peripherals. The profile specifies procedures on a group of devices, while other profiles, such as BAP and VCP, define point-to-point procedures. CAP defines procedures on a group of devices for both Unicast and Broadcast as defined in the BAP profile. CAP defines procedures for a group of devices for Volume and Microphone control as defined in the VCP and MICP profiles.

The Content control services, such as call and media, are referencing CAP to provide context to deploy on multiple devices via CCIDs lists and Context Types mapping to the Audio Stream. The Content control services have a server functionality, which serves multiple devices, and therefore, the coordination and synchronization across groups of devices is naturally defined in the content control profiles.

Table 5.2 describes the CAP roles and their functionality.

TABLE 5.2

CAP Roles and Functionality

Role	Description of Functionality	Examples
Initiator	The Initiator starts, updates, or ends Unicast or Broadcast Audio Streams with one or multiple Acceptors. It can expose BASEs. It provides information on the use case context of Unicast or Broadcast Audio Streams as described by Context Type values. It provides information on the association of Unicast or Broadcast Audio Streams with Content Control services through CCIDs. It can be a set coordinator and discover Acceptors that are members of a Coordinated Set.	A phone, PC, and television are examples of devices that would implement the Initiator role.

(Continued)

TABLE 5.2 (*Continued*)
CAP Roles and Functionality

Role	Description of Functionality	Examples
Acceptor	The Acceptor can expose Audio Stream Endpoints (ASEs). It can receive Broadcast Audio Streams transmitted by an Initiator. It receives information on the use case context of Unicast or Broadcast Audio Streams as described by Context Type values. It receives information on the association of Unicast or Broadcast Audio Streams with Content Control services through CCIDs. It can delegate scanning for Broadcast Audio Streams to a Commander acting as a BAP Broadcast Assistant. It can receive requests to start or end reception of Broadcast Audio Streams from a Commander acting as a BAP Broadcast Assistant. It can adjust the volume of audio rendered based on requests from a Commander. It can mute and/or set the signal level of its microphone based on requests from a Commander. It can host clients who use a Content Control Service. It can be a set member in a coordinated set.	A headset, earbud, hearing aid, and loudspeaker are examples of devices that would implement the Acceptor role.
Commander	The Commander can scan for Broadcast Audio Streams and their associated metadata on behalf of an Acceptor. It can request Acceptors to start or end reception of Broadcast Audio Streams transmitted by an Initiator and provides information on their use case context as described by Context Type values. It can control the volume of the audio rendered by Acceptors. It can control the mute state and/or signal level of a microphone on Acceptors. It is a set coordinator and discovers Acceptors that are members of a Coordinated Set. It can host clients who use a Content Control Service.	A remote control used to adjust the volume of a pair of hearing aids is an example of a device that would implement a stand-alone Commander. Incorporated in such a remote control is likely also the ability to play/pause media. A Commander may be colocated with an Initiator in case the Initiator is a source of a Broadcast and also configure an Acceptor to sync to the Broadcasted content.

The following procedures are supported by CAP; each of the procedures uses underlying BAP or VCP procedures on multiple audio peripheral server devices:

- Unicast audio starting procedure
- Unicast audio updating procedure
- Unicast audio ending procedure
- Broadcast audio starting procedure
- Broadcast audio updating procedure
- Broadcast audio ending procedure

- Broadcast audio reception start procedure
- Broadcast audio reception ending procedure
- Unicast to Broadcast Audio Handover procedure
- Broadcast to Unicast audio Handover procedure
- Distribute Broadcast_Code procedure
- Change volume procedure
- Change Volume Offset procedure
- Change Volume Mute State procedure
- Microphone Mute State procedure
- Change Microphone Gain Setting procedure
- Find Content Control Service procedure
- Coordinated Set Member Discovery procedure

The Unicast/Broadcast start/stop/update are done by the Initiator device to begin an audio use case, update its context, and end the audio use case across multiple Acceptors in a uniform manner. The Broadcast audio reception start/ending is done by the Commander role to initiate or stop reception of Broadcast by multiple Acceptor devices in a uniform manner. The volume and input gain procedures are done by the Initiator, or by a Commander, to control speakers and microphones in multiple Acceptors in a uniform manner.

REFERENCES

1. Bluetooth Core 6.0 or later, https://www.bluetooth.com/specifications/specs/core-specification-6-0/
2. BAP version 1.0.2 or later, https://www.bluetooth.com/specifications/specs/basic-audio-profile-1-0-2/
3. PACS version 1.02 or later, https://www.bluetooth.com/specifications/specs/published-audio-capabilities-service-1-0-2/
4. ASCS version 1.0.1 or later, https://www.bluetooth.com/specifications/specs/audio-stream-control-service-1-0-1/
5. BASS version 1.0 or later, https://www.bluetooth.com/specifications/specs/broadcast-audio-scan-service/
6. VCP version 1.0 or later, https://www.bluetooth.com/specifications/specs/volume-control-profile-1-0/
7. VCS version 1.0.1 or later, https://www.bluetooth.com/specifications/specs/volume-control-service-1-0-1/
8. VOCS version 1.0.1 or later, https://www.bluetooth.com/specifications/specs/volume-offset-control-service-1-0-1/
9. AICS version 1.0 or later, https://www.bluetooth.com/specifications/specs/audio-input-control-service-1-0/
10. MICP version 1.0 or later, https://www.bluetooth.com/specifications/specs/microphone-control-profile-1-0/
11. MICS version 1.0 or later, https://www.bluetooth.com/specifications/specs/microphone-control-service-1-0/
12. CCP version 1.0 or later, https://www.bluetooth.com/specifications/specs/call-control-profile-1-0/

13. TBS version 1.0 or later, https://www.bluetooth.com/specifications/specs/telephone-bearer-service-1-0/

14. MCP version 1.0 or later, https://www.bluetooth.com/specifications/specs/media-control-profile/

15. MCS version 1.0.1 or later, https://www.bluetooth.com/specifications/specs/media-control-service-1-0-1/

16. CSIP version 1.0.1 or later, https://www.bluetooth.com/specifications/specs/coordinated-set-identification-profile-1-0-1/

17. CSIS version 1.0.1 or later, https://www.bluetooth.com/specifications/specs/coordinated-set-identification-service-1-0-1/

18. CAP version 1.0 or later, https://www.bluetooth.com/specifications/specs/common-audio-profile-1-0/

19. CAS version 1.0 or later, https://www.bluetooth.com/specifications/specs/common-audio-service-1-0/

20. LC3 version 1.0.1 or later, https://www.bluetooth.com/specifications/specs/low-complexity-communication-codec-1-0-1/

6 LC3 Codec

In this chapter, we will look into the details of the mandatory LE Audio Codec – namely, LC3 (Low Complexity Communication Codec). The LC3 Codec is the main Codec used in LE Audio. Other optional (externally defined) or vendor-specific codecs may also be used. In this chapter, the LC3 Codec is reviewed from various perspectives. An algorithmic overview of the Codec is provided, and various aspects of the Codec features are outlined, such as compression quality, latency and complexity, the overall system delay, and bitrates.

WHY A NEW CODEC?

We already mentioned the need to use audio codecs in general, and for Bluetooth® audio in particular. Using codecs allows the audio applications to compress the digital audio content into smaller bitrates for transmission over the air. Bluetooth audio compression is essential due to the requirement to send low user-bitrates over the radio. In LE Audio, compression is more essential in order to free more spectrum and enable low-energy operation as well as new use cases. As we discussed in the previous chapters, LE Audio enables multistream use cases, and providing lower bitrates enables sending Audio Streams in a multi-device topology.

The previous mandatory codecs used by Bluetooth® Classic provided a certain level of compression and quality that was considered good at that time. The main Codec is SBC for music use cases and its mSBC variant for voice use cases. It was later shown that SBC was considered to provide a medium music listening quality, and other optional codecs and vendor-specific codecs began to emerge in order to provide better listening quality. Optional codecs were added to Classic Audio music profiles, while vendor-specific codecs were used by proprietary applications. The other optional and vendor-specific codecs required higher bitrates.

In recent years, a lot of progress has been made in acoustic and compression technologies. The result was codecs that are using better time to frequency transformations. These codecs had the potential to produce a lower bitrate and better quality compared to SBC. The new codecs, however, were more complex in terms of the amount of memory and CPU or DSP cycles required for encoding and decoding. For example, in the cellular industry, there are advanced codecs that reside in cellular base station towers and in handset cellular devices (phones). The cellular codecs use the latest state-of-the-art compression and decompression technology to provide excellent sound quality to mobile phone users.

In many cases, the codecs in cellular technology use proprietary technology and require royalty payments. These codecs are also highly complex and require strong processing power and large amounts of memory. In the cellular world, phones today are actually a small PC and do have the processing power to run such codecs. The same goes for the base station towers, which may employ high processing units for

DOI: 10.1201/9781003590187-6

compression and decompression logic. This is true for cellular radios between base stations and smartphones.

In Bluetooth technology, however, there is another category of small devices, such as hearing aids, small speakers, microphones, and earbud devices. The category of small Bluetooth peripherals requires the Codec to be less complex in order to allow the peripheral devices to consume less power and run longer on battery charging.

Due to the preceding reasons, the Bluetooth SIG aimed to develop a new Codec which will provide the latest state-of-the-art compression and decompression, better listening quality, and latency, and yet employ lower complexity compared to the other advanced codecs from the same category. The Codec, which was developed for this purpose, is LC3, which indeed stands for Low Complexity Communication Codec.

Figure 6.1 shows the high-level flow of the LC3 Codec when used over the LE Audio Stream.

LC3 operation from source capturing device to sink playback device

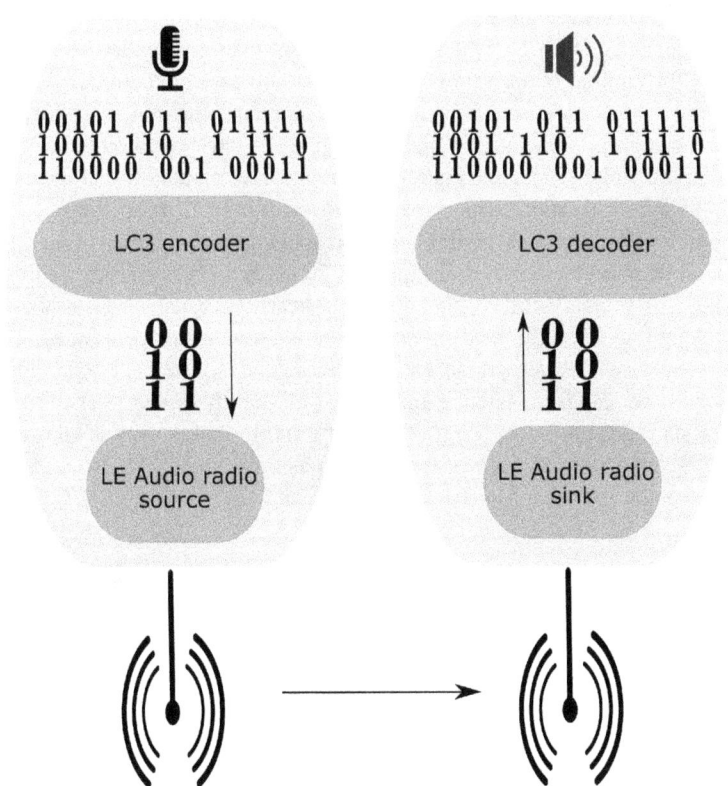

FIGURE 6.1   LC3 Codec operation.

As shown in Figure 6.1, the source of audio is generating audio samples as a bit stream, which is fed into the LC3 encoder part. The source audio samples are uncompressed PCM samples. For example, 48 kHz generates 480 samples every 10 ms. Each sample may be a 16-bit signed number, which represents the signal amplitude over time. The output of the encoder part generates a lower compressed bit stream that is sent over the air to the remote device. The compression ratio of LC3 is about eight, and it is better than the legacy SBC compression. The remote device is the sink of the audio and is used for playback. The receiving device receives a compressed bit stream from the LE Audio radio and passes the bit stream into the decoder. The decoder handles the incoming bit stream and decodes it to recover the audio samples as close as possible to the originally sampled at the source device. At this point, the audio samples may be played on a speaker. In music use cases, one device acts as a source, and another device acts as a sink. In bidirectional communication, such as voice or hybrid use cases, both devices act as a source and a sink and employ both the LC3 encoder and decoder parts.

LC3 COMPARED WITH SBC/MSBC

LC3 provides better listening quality compared with SBC and produces a smaller bitrate. Since LC3 uses the latest audio Codec technology, it is more complex compared with SBC; however, LC3 complexity is lower when compared to other state-of-the-art codecs (e.g., OPUS), which provide the same audio quality. Complexity is measured by the number of operations required to compress or decompress the bitrate. Another aspect of LC3 is that it produces a lower latency compared with other codecs in the same category. Low-latency operation in Bluetooth® technology is essential since it is the last link in the audio chain. Audio may have an end-to-end latency from the source to the destination, which may begin from cellular base stations or from Internet audio streaming. In particular, for voice applications such as cellular, the margin of Bluetooth audio latency is in the order of 20 ms, since it is already adding up to other latency components of the cellular call. In hearing aid use cases, it is essential to keep latency low since the hearing aid device is often amplifying ambient sound from a close-by wireless microphone. LC3 is meeting the requirement to provide voice coding in the order of 20 ms latency and similar or slightly higher latency for music use cases when required. In certain use cases, the end-to-end latency is allowed to be higher in order to enable audio processing, such as noise reduction, before the playback. Table 6.1 summarizes the comparison between LC3 and SBC/mSBC.

TABLE 6.1
LC3 to SBC Comparison

Property	SBC/mSBC	LC3
Applications	Music and voice	Music and voice
Quality	Medium	High
Latency	Low	Low
Complexity	Lower	Low
Bitrate	High	Low

In the next few sections, we will review each category from Table 6.1 in more detail.

LISTENING QUALITY

As part of the LC3 Codec development, the Bluetooth® SIG conducted subjective listening tests to assess the quality of the LC3 Codec as well as objective measurements such as PEAQ for music (Perceptual Evaluation of Audio Quality) and POLQA for speech (Perceptual Objective Listening Quality Analysis).

The results showed that voice LC3 has the same voice quality at half the bitrate when compared to mSBC. And in the case of music, the LC3 quality is always better than SBC, at about half the bitrate when compared to SBC high-quality settings.

Figure 6.2 shows a music quality comparison between the SBC high-quality setting and the LC3 music setting. As shown, the quality of LC3 is close to the original reference audio file when no compression is done. In the quality comparison, LC3 uses half the bitrate of SBC.

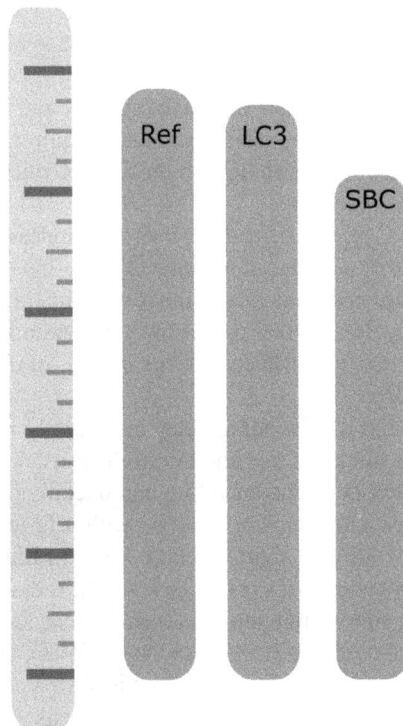

FIGURE 6.2 LC3 music quality.

In subjective listening tests, a group of expert listeners listen to various audio samples and score the experience on a scale of 1 to 5. The listeners do not know if the audio is encoded and what Codec is used, if any. There is a hidden reference file that is uncompressed within the various files. The listeners score the reference file as well as the compressed files based on subjective listening only. In various listening tests done by expert listeners, LC3 consistently scored close to 5, while the SBC quality is closer to 4, at almost double the bitrate compared to LC3. A score closer to 5 is defined as imperceptible, which means that it is almost impossible to tell the coded/decoded samples from the original soundtrack, while a score of 4 is defined as perceptible, in which a difference in quality is perceived. A score closer to 3 is defined as slightly annoying, with annoyance increasing when the score is 2 or 1. The listening test procedure described earlier is also known as MUSHRA (MUltiple Stimuli with Hidden Reference and Anchor).

LATENCY AND COMPLEXITY

A Codec adds latency to the audio chain. In general, audio streaming or voice calls already contain latency elements regardless of Bluetooth® audio. For example, listening to audio streaming over the Internet includes a network delay. Another example is a voice call, in which the cellular network and cellular codecs introduce delay. As mentioned earlier, the gateway source of audio may also be using codecs. In the case of Bluetooth technology, there is a transcoding operation in which audio is recovered from the network or cellular coding and then compressed again for transmission over the Bluetooth connection.

As Bluetooth technology is the last link to the speaker and microphone, it is essential to keep its latency as low as possible. The latency requirements may be relaxed in the case of music playback and more tight in the case of voice or low-latency media such as gaming or movie playback.

For the Bluetooth latency, we defined the overall Wireless System Delay (WSD). The WSD consists of various latency components. It begins with the time it takes to capture an audio frame for encoding. It continues with the time it takes to encode the audio frame and then the time it takes to transport the audio frame over the air from the Bluetooth source device to the Bluetooth sink device. At the sink device, the LC3 decoding delay is added.

The encoding delay and the decoding delay are directly impacted by the Codec complexity for the given sampling rate and bitrate. These delays are impacted by the number of required operations. The encoding and decoding delays also depend on the implementation clock rate. Implementations with a faster clock rate will complete the encoding or decoding in shorter delays. Since in Bluetooth the LC3 Codec also resides in small devices with a slower clock rate, it is essential to optimize LC3 complexity accordingly, so it would allow reasonable delays also in small devices.

There is an additional component that is added to WSD, and it is called the presentation delay. The presentation delay placement is always at the peripheral device. In the case that the peripheral device is a wireless speaker, then the presentation delay is added after the decoding and includes the decoding time along with other components such as DAC delays (Digital to Analog Conversion), noise reduction elements, and jitter buffers. In the case that the peripheral device is a wireless microphone, then the presentation delay is added before the encoding and includes the encoding time along with other components such as ADC delays (Analog to Digital Conversion) and echo cancellation elements. The purpose of the presentation delay is to allow additional audio processing at the peripheral side. In the case of multi-device topology, the presentation delay is set equal among all peripheral devices, so that it is the lower common denominator of the best-case delay among all peripheral devices. The agreed value is communicated to all peripheral devices by BAP (see Chapter 5). This allows all peripherals to synchronize the audio, such that audio is rendered at the same time on multiple speaker devices or acquired over the air at the same time from multiple microphone devices. The presentation delay for microphones is also known as the acquisition delay. The presentation delay for speakers is also known as the rendering delay. The additional presentation delay on top of the encoder or decoder depends on the application. As an example, a presentation delay may be as low as 1 ms or as high as 60 ms. The peripheral reports a min and max range, and BAP mandates that 40 ms will be included in the range to assure minimum interoperability. As an example in a two-peripheral topology, if Peripheral A reports min = 4 ms and max = 50 ms and Peripheral B reports min = 2 ms and max = 45 ms, then the selected presentation delay for both Peripherals A and B will be configured to be 4 ms, since this is the minimum presentation delay that is supported by both peripherals. Although Peripheral B supports a minimum of 2 ms, this delay is not supported by Peripheral B, so the BAP Client selects the value of 4 ms, which is the maximum between the minimums of the two devices.

The LC3 Codec works on units of frames. The frame duration is a configuration value of LC3. There are two frame duration modes possible in LC3 for LE Audio: 7.5 ms and 10 ms. In each mode, a full frame is processed and encoded after capturing. In addition to a full frame delay, LC3 also generates an inherent look-ahead algorithmic delay. The encoder works on two frames in order to compress the next frame. A portion of the previous frame audio content is encoded in the current frame. And the audio content tail of the current capture is saved and encoded in the next frame. This property of LC3 is called LD-MDCT (Low Delay Modified Discrete Cosine Transform). The cosine transform serves as the filter when transferring time domain samples to frequency domain samples. And the width of the cosine filter is over two frame durations. As a result of LD-MDCT, a look-ahead delay of 4 ms is added to the 7.5 ms frame, and a look-ahead delay of 2.5 ms is added to the 10 ms frame. The look-ahead delay is algorithmic only and represents a delay in audio content, and not actual processing time. This delay is added to WSD since it adds to the end-to-end latency of the audio content.

The various components of LC3 latency are shown in Table 6.2 for the two LC3 modes, 7.5 and 10, with examples for estimated values for the main components. An actual implementation may have smaller or higher values depending on the processing clock. Not shown in the table are elements for jitter, buffering, DAC, and ADC, which also depend on actual implementation. The actual values for the encoder and the decoder will be different for different sampling rates. The values in the table for encoding and decoding provide a ballpark figure for the LC3 Codec latency. The Bluetooth® LE transport delay in this example assumes minimum transport latency, within a single transport interval, which is common in voice use cases; music use cases may use longer transport latency to accommodate more retry opportunities for higher reliability.

TABLE 6.2 Example for LC3 WSD Latency Components

Delay Component Breakdown	LC3 7.5 [Total Latency: 16 ms + presentation delay] [ms]	LC3 10 [Total Latency: 18.4 ms + presentation delay] [ms]
Look-ahead	4	2.5
Capturing	7.5	10
Encoding (Plus acquisition for peripheral)	2.2	2.8
Bluetooth® LE Transport	1.1	1.3
Decoding (Plus rendering for peripheral)	1.4	1.8
Additional Presentation	1 to 60	1 to 60

Figure 6.3 shows the delay component contribution over a timeline. The LD-MDCT transformation filter is shown at the top. The LD-MDCT transformation is done at the encoder, and it uses samples from the previous captured frame (N−1). Overall, the encoder works on a double window relative to the capturing window. The inverse LD-MDCT is shown at the bottom of the figure. The inverse LD-MDCT is used by the decoder, and it is using previous frame decoded information when decoding the current frame. Some of the decoded information (the tail) is saved for the next decoded frame (N+1). The encoding and decoding are using LD-MDCT, and inverse LD-MDCT enhances the recovery back to the original audio content by using samples from previous frames. This property helps to smooth sharp tone transitions in certain tracks in the sound. This is the LC3 property, which provides close transparency audio recovery. This algorithm introduces the look-ahead delay, so audio content is delayed by a partial frame, in addition to the capturing frame delay. Acquisition and rendering are delays that depend on the peripheral configuration for either a microphone or a speaker. In many cases, only one of these delay components is valid.

The example in Figure 6.3 shows a low-latency application where audio is rendered immediately as it is available. In this case, the Bluetooth LE transport is configured to deliver packets within the current frame only. The figure shows how the total frame N latency is roughly around two times the LC3 frame size (capturing window). There are applications that may allow longer latency for rendering. For example, music playback may allow longer latency in order to achieve higher reliability. In this case, the Bluetooth LE transport may be longer in cases where packets need more retransmissions for a longer period of time. We will get back to this aspect when discussing the link layer transport in Chapter 7.

MDCT tranform

look ahead   N-1   N   N+1

capturing   • • • frame N-1 frame N   frame N+1   • • •

acquisition

encoding

LE transport

decoding

rendering

inverse MDCT tranform

<-- frame N latency -->

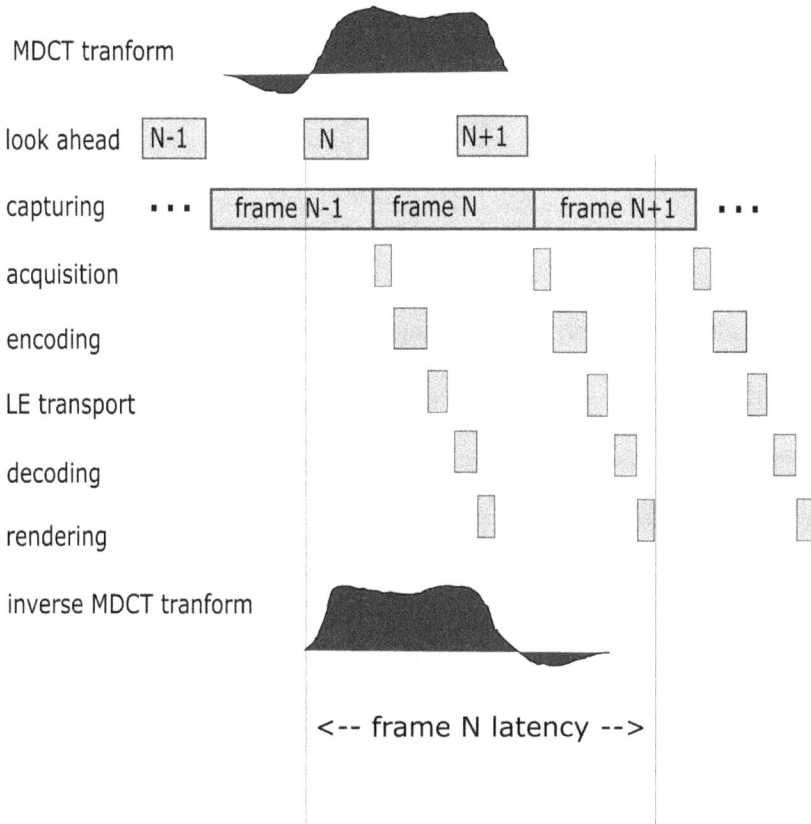

FIGURE 6.3   LC3 WSD components.

BIT RATE

LC3 spans a full range of sampling rates for voice and audio.

Among the various configuration options, there are two main differentiating vectors. Voice and music are the first vector. The hearing aid quality or high quality audio is the second vector.

The main configurations for hearing aid applications are sampling rates of 16 kHz for voice and 24 kHz for music. In the first chapter, we reviewed the hearing aid requirements for people with hearing loss and mentioned that higher tones than 11 kHz are not audible by hearing aid users. This is why the music sampling rate stops at 24 kHz (sampling rate is twice the tone content). Additional audio processing for hearing aids takes care to replace higher tones with tones under 11 kHz, which are covered by the 24 kHz sampling rate setting. The voice quality is a wideband speech (WB) and may be used by hearing aid users as well as people with no hearing loss.

For a high-quality voice application, an additional configuration is added: 32 kHz, which adds super wideband speech (SWB). With SWB, the voice quality adds an in-room experience, where, during a voice call, the user may hear the call as if the user is within the same room as the counterpart on the other end of the call. This configuration has already become popular in cellular phones, and Bluetooth® technology may now provide the same level of user experience.

For high-quality music or hybrid voice and music, 48 kHz full band is available. 48 kHz provides the full range of the human ear's audibility, which is sensitive to tones of up to 20 kHz, as we saw in Chapter 1. LC3 is designed as a mono Codec, so that left and right stereo may be encoded separately. This allows sending left and right streams to two separate speakers or earbuds.

Table 6.3 summarizes the bitrate and payload size that is required for each LC3 frame duration and application configuration. The table summarizes the most common configurations. There are other possible LC3 sampling rates, such as 8 kHz for voice and 44.1 kHz for music, which are supported by LC3 but are less commonly used. There are also different bitrates available in LC3. For example, 48 kHz may be used with a bitrate of 124 kbps instead of 96 kbps for a marginal increase in quality. Table 6.3 shows the configuration for a single channel of audio, such as mono, left side, or right side. Stereo requires twice the bitrate.

TABLE 6.3
LC3 Bitrates and Payload Sizes per Sampling Rate

Application	Sampling Rate [kHz]	LC3 7.5 Bitrate/Payload [Kb/s]/[bytes]	LC 10 Bitrate/Payload [Kb/s]/[bytes]
Hearing aid voice	16	32/30	32/40
Hearing aid music	24	48/45	48/60
High-quality voice	32	64/60	64/80
High-quality music	48	96/90	96/120

REFERENCE

1. LC3 version 1.0.1 or later, https://www.bluetooth.com/specifications/specs/low-complexity-communication-codec-1-0-1/

7 Transport Layer

In this chapter, we will study the details of the Core Controller in LE Audio, which constitutes the LE Audio transport. We will review the LE Core features that enable LE Audio and the changes to the Bluetooth® Core Specification that further enhance the LE Audio experience. The LE Audio transport supports the Host BAP procedures to create Audio Streams over LE.

ISOCHRONOUS CHANNELS

Isochronous Channels are located at the heart of the LE Audio Core Controller. The Isochronous Channels are a set of user channels, logical links, logical transports, physical links, and physical transports. The Isochronous Channels enable the setting up of audio over LE. Before going into the details of Isochronous Channels, let's review the application requirements that are driving the Isochronous Channels.

AUDIO QUALITY OF SERVICE

One of the key requirements for the audio transport is time-bound data, which is transmitted and received in relation to an Audio Source that is producing the audio as digital data frames at a certain quality level. This requirement translates to an application QoS. The QoS types for audio are

- Latency
- Target bitrate
- Reliability

LATENCY

Audio applications require the data to be delivered from the source to the sink within a given amount of time.

The audio is produced as compressed audio frames, and it has a limited lifetime. Unlike other data, such as files or phone book entries, audio should reach the destination application within a certain amount of delay. A user of an audio application will experience poor listening quality if the audio is delayed beyond a certain threshold. A large delay in audio may cause different unwanted effects, such as missing audio sections, echo, and sound that is not synchronized. A large delay will cause audio to reach later than the listening user learns to expect it. Therefore, the delay is bounded such that the transport has a maximum limit of how much it can delay a given audio SDU.

DOI: 10.1201/9781003590187-7

BIT RATE

The bitrate targets are higher for higher-quality audio. In Chapter 6, about the LC3 Codec, we reviewed the various audio quality configurations and their required bitrate and frame payload sizes per frame interval. A certain audio quality requires a certain bitrate. The application bitrate is derived from the frame size and frame duration for a given audio bandwidth. The audio bandwidth results in a certain compression ratio of the data. The audio is sampled at fixed intervals, and then frames are compressed into SDUs. As a result, the application is producing X number of bytes every Y number of ms. For example, one of the LC3 Codec high-quality settings produces 240 bytes every 10 ms for a two-channel stereo, 120 bytes for the left channel, and 120 bytes for the right channel. This translates into a bitrate of 192 kbps. Another LC3 setting produces 180 bytes every 7.5 ms, for the same effective bitrate of 192 kbps (90 bytes for the left channel and 90 bytes for the right channel).

RELIABILITY

Audio applications require a certain amount of reliability, so that audio content integrity is guaranteed in poor link quality radio conditions (e.g., longer distance between the source and the sink) or in congested environments with multiple radios, which increases chances for packet collisions.

In other applications, such as in the example of files or phone book entries, the data is known to be of the best-effort QoS type. When sending traffic over a Bluetooth® link, packets of data may be corrupted due to the wireless link quality or a congested environment with multiple transmitters. In this case, the packet of data is retransmitted. With best-effort QoS, the retransmission has no limit, since the only requirements are reliability and integrity of the data. In the best-effort case, packets may be retransmitted many times, until successfully received by the peer device, or the link is dropped in case of getting out of reception range or if the interference is completely blocking the signal.

On the other hand, audio has a target bitrate QoS and a low-latency QoS. Audio data requires a different behavior from the Bluetooth® LE link layer. Bluetooth LE data for audio may not be retransmitted beyond a certain limit. A reliability setting determines the number of times an audio SDU is allowed to be retransmitted. The number of retransmissions depends on the type of audio application. Applications that are interactive limit the retry to a small window, while music listening may use a larger retry window to prevent loss of data, while an application player may compensate for the delay by buffering and controlling the playback speed, and the sink may maintain a jitter buffer to make sure samples are available to render also when over-the-air data is delayed due to temporary interference.

Reliability translates to the number of retransmissions to ensure that the audio packet arrives at the receiver in case a radio interference corrupted a previous packet. The QoS for audio is often a trade-off between reliability, latency, and bitrate. For example, in media audio applications, reliability and bitrate are given higher priority, which translates into longer latency and more retransmissions, while in voice audio applications, low latency is given higher priority, which requires smaller bitrate and less retransmissions, so reliability is given less priority, and integrity of data is achieved via other methods such as playing the previous audio samples in case audio content is corrupted by the radio link quality.

This latter method is known as PLC (Packet Loss Concealment); in PLC, an algorithm determines how to complete lost audio content when the radio is less reliable (fewer retransmissions). The PLC objective is to prevent clicks and pop noise that happens when audio samples are missing. There are simple PLC algorithms and more advanced PLC algorithms. PLC is more common in voice audio applications due to the interactive nature and low latency, which forces the usage of small retransmission windows. PLC may also apply to media/music use cases with low latency, such as gaming or online interactive music jam.

ISOCHRONOUS CHANNELS TYPES

As we saw in the previous chapters, there are two main types of LE Audio topologies: Unicast audio and Broadcast audio (also known as Auracast™ broadcast audio). Unicast audio serves point-to-point topologies or point-to-multipoint topologies with link quality receiver feedback over bidirectional links. Broadcast Audio's goal is to share audio to an unlimited number of listeners in a connectionless unidirectional link, with no link quality feedback from the receivers.

Isochronous Channels provide two logical transports to support Broadcast audio and Unicast audio:

- CIS: Connected Isochronous Stream
- BIS: Broadcast Isochronous Stream

CIS LOGICAL TRANSPORT

When two Bluetooth® LE devices are connected, the device with the Central role in the connection may create CIS, by initiating a request to the device with the Peripheral role. For example, the Central device may be a phone or a PC, and the Peripheral device may be an earbud or a headset. If the Central is connected to more than one Peripheral audio device, then the Central may initiate the creation of CIS with each Peripheral. An example is a phone creating one CIS with the left earbud and another CIS with the right earbud. A CIS is always part of a group, which is called CIG (Connected Isochronous Group). The CIG can contain one or more CISes to different Peripherals. The owner of the CIG is always the Bluetooth LE Central device. And it is the responsibility of the Central device to create all CISes in a CIG in a synchronized manner, such that the Peripherals are synchronized to the Audio Source or producing audio in synchronization with each other.

The flow of events is that the BAP profile negotiates the creation of ASE records between the BAP Client and the BAP servers. When the ASE records are configured with Codec settings and QoS settings, and while enabling the ASE records, the BAP Client instructs the link layer to create a CIG and then create CIS with each Peripheral (see Chapter 5 for more details about BAP).

Each ASE record is a flow of audio in a given direction. ASE records of opposite directions to/from the same device are mapped to a single CIS at the logical transport. CIS allows streaming of data in both directions, and it is a logical transport for Sink ASE and Source ASE to/from the same Peripheral. ASE to different devices is mapped to different CISes, since each CIS is created between two Bluetooth LE devices. For example, a BAP Client in a phone connected to two BAP servers in two earbuds will create a single CIG followed by two CISes, one for each earbud. If only the right earbud has a microphone, then the two right earbuds' ASEs, source ASE and sink ASE, map to one CIS to that right earbud Peripheral, while the left earbud sink ASE will map to another CIS to the left earbud Peripheral. Note that the CIS that maps to the single sink ASE (left audio) is still a bidirectional connection, and in this case, the packets in the reverse direction from the Peripheral to the Central are empty packets carrying acknowledgments as link quality feedback. Based on these acknowledgments or lack of acknowledgments, the Central knows if a given CIS PDU requires a retransmission. Acknowledgments are also sent in bidirectional CIS when audio is sent in both directions, as acknowledgment info is placed in the PDU header as logical link information. CIS PDUs that are sent to carry audio data may also carry acknowledgments of audio data that was received in a previous CIS PDU in the reverse direction on the same CIS.

The CIS logical transport moves any data that the application may send according to the QoS parameters, which were set for latency, bitrate, and reliability. The data from the application is sent to ISOAL channels. The ISOAL channels allow the application to send time-bound data such that it is adapted from application SDU to link layer CIS PDUs. ISOAL provides decoupling of the application QoS settings from the over-the-air PDU parameters. The scheduler in the controller is responsible for sending the data at the rate and timing that the application requires.

In stereo music use cases, LC3 is producing compressed SDUs for two encoded channels: left and right. In the case that the left and right channels are targeting two peripherals, each CIS will carry SDUs from a different channel. While sending audio data to a single peripheral, such that in the case that the Central connects to a single headset device, the Central can multiplex the two left and right channels into a single ASE and generate SDUs with two channels, left followed by right. In this case, the CIG will contain a single CIS, and each CIS PDU may carry LC3 encoded left audio data followed by LC3 encoded right audio data. At the headset, the two channels are separated from the SDU (left followed by right) and decoded separately and rendered on two speakers (one on each ear). Alternatively, the stereo music stream may be encoded by LC3 as a mono stream with a single-channel SDU, with no stereo effect. Certain implementations may prefer to configure two ASEs to the same peripheral for the headset case, left ASE and right ASE, which will require setting a CIG with two CISes to the same peripheral, one CIS to carry left audio and another CIS to carry right audio. In this case, separate SDUs are sent for the left channel and right channel, which are sent to the same device over two separate CISes.

Figure 7.1 shows an example of two CIS logical transports used by a Central device when communicating with two Peripheral devices. In this example, the Central is carrying left and right stereo music streams such that each audio direction is sent to a different Peripheral device. The Peripheral devices may be a pair of speakers. In this case, no audio is sent from Peripherals to the Central. Other use cases such as voice call use cases or voice over IP also allow Peripherals to send back audio data from a microphone. Figure 7.1 shows an example in which the Central connects to two different devices for left stereo and right stereo and uses LC3 encoded data for the left channel, separate from LC3 encoded data for the right channel.

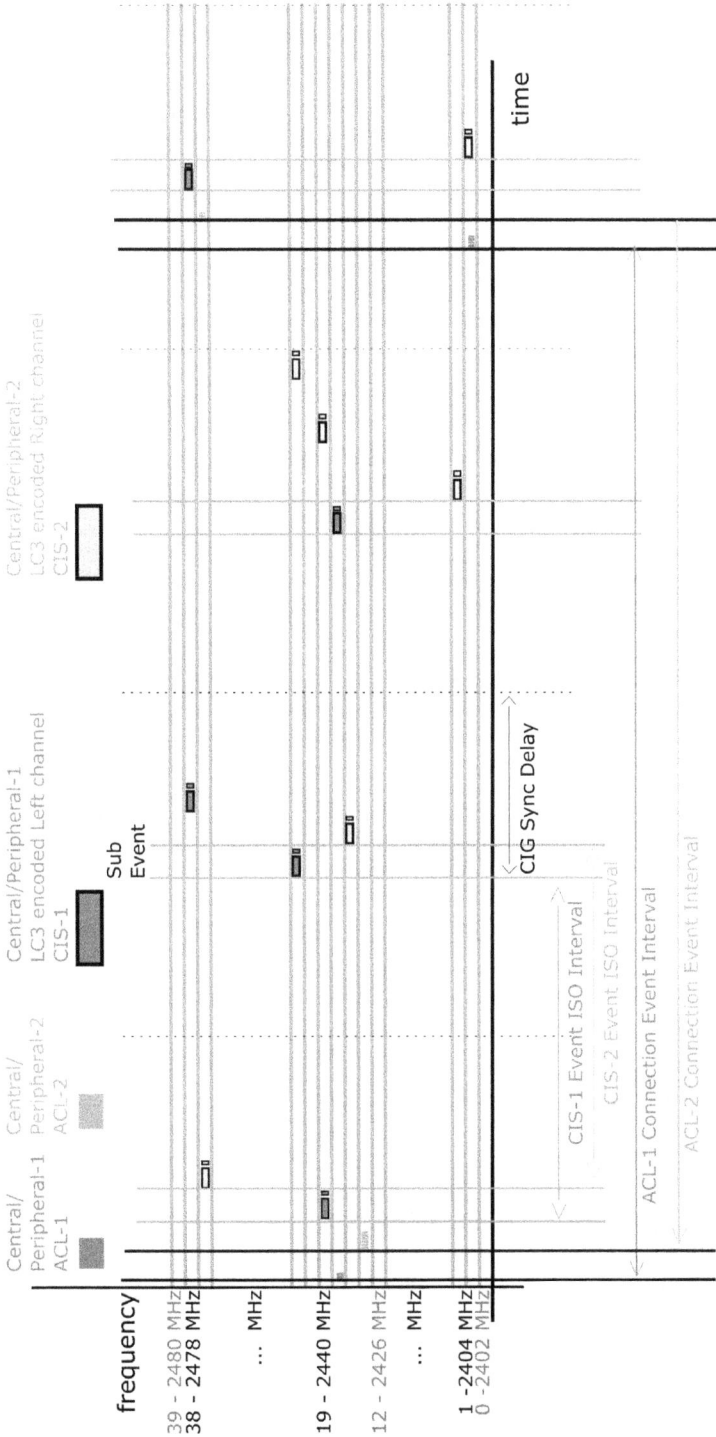

FIGURE 7.1 CIS Events over time and frequency.

In Bluetooth LE, there are 37 data channels, while channel indexes 0, 12, and 39 are reserved as primary Advertising channels. The example shows how the central is assigning the data channels over time. The central maintains the ACL logical transport with each of the peripherals. The ACL logical transport is used for managing the connection by the link layer (e.g., radio close loop power control procedures) and carrying Control layer messages from Host profile applications as L2CAP/GATT data (e.g., changing volume procedures). As seen in Figure 7.1, the ACL connection intervals maintain a longer interval compared to the CIS event intervals, which are carrying audio data. ACL packets are typically shorter since they are carrying profile signaling data, while CIS data packets are longer as they are carrying encoded audio. Another difference is that each ACL Connection Event uses the same RF channel for all transmissions within this event, while each CIS event is subdivided into CIS subevents, each on a different RF channel. The concept of CIS subevents allows using a different channel for each subevent. A CIS subevent contains two PDUs, the first PDU from the Central to the Peripheral, followed by a PDU from the Peripheral to the Central. The default inter frame space between consecutive PDUs is 150 μs in both ACL and CIS, and the value can be negotiated to a lower duration. In this example, the communication is unidirectional music streaming, so the CIS PDU from the Central is larger, while each Peripheral responds with an empty packet containing an acknowledgment message. In other use cases, such as voice calls, the return packets from the Peripheral may carry audio data in addition to the acknowledgment on the speaker audio data; similarly, the Central will use the same PDUs to carry speaker data and acknowledgment on the microphone audio data from a Peripheral.

The example in Figure 7.1 focuses on music streaming. The CIS event may contain more than one subevent for each Peripheral. The purpose of the subevents could be to carry different parts of the audio content in multiple burst numbers (BN) or to allow for retry. In this example, subevents are used for retry, so BN = 1 and the number of subevents (NSE) is three for each CIS, and all the subevents are used for retry. In this example, the subevents are scheduled by the Central such that they are interleaved between the Peripherals. Subevent 1 is sent to Peripheral 1, followed by Subevent 1 to Peripheral 2, and then Subevent 2 is sent to Peripheral 1, followed by Subevent 2 sent to Peripheral 2, and so forth. While subevents are used for retry, additional subevents are only present in case the PDU fails to transmit successfully. The example in Figure 7.1 shows such cases. For example, the third CIS event to Peripheral 2 required three subevents until the CIS PDU was acknowledged by the Peripheral; while in the same CIG event, the CIS event to Peripheral 1 required only a single subevent to deliver the PDU. The CIG Sync Delay is defined as the overall time to deliver CIS PDU to all CIS in a CIG. The CIG event is the superset of all CIS events as configured by the application for a given use case. In our example, the application is stereo music, and we have a CIG event at each interval, which contains two CIS events, one CIS event for each Peripheral (left device and right device). In the example in Figure 7.1, the number of subevents to each Peripheral is defined as a maximum of three. Each CIG event has one original content plus two retry opportunities for each Peripheral. In this case, both Peripherals will use the CIG Sync Delay after three subevents, regardless of how many retries were actually used. This allows left and right content to render on both speakers at the same time. Note that there is an additional presentation delay configured by BAP on top of the synchronization point to take care of decoding delays (see Chapters 5 and 6).

In music use cases, where a larger delay is allowed by the application, it is also common to add an overall longer transport delay on top of the CIG Sync Delay. The longer transport delay is configured by the Central device, and will allow it to continue retrying the content also in the next CIS events. In the link layer, these parameters translate into flush timeout (FT). FT defines additional intervals to continue the retransmissions, and it provides time diversity and improves the reliability of the stream. So, for example, if the number of subevents is 3 and FT is 6, then there are overall 18 transmissions possible for the same content: 1 original content and 17 possible retransmissions. In this case, the synchronization point to render audio takes into account also the transport delay as the worst case. In the example of FT equals 6 and CIS ISO interval equals 10 ms, then the overall transport latency would be 50 ms plus CIG Sync Delay; since the last interval opportunity ends at the CIG Sync Delay (sample calculation: {10 ms x [6−1]} + CIG Sync Delay). This value will be the synchronization reference point used by the application before applying a presentation delay. In voice use cases and interactive use cases, FT is set to 1, and the transport latency consists of the CIG Sync Delay alone. There are also low-latency music use cases, such as gaming music/media or movie music/media, where the transport latency will be in a low to mid-range, so FT may be 1 or 2. For example, if FT equals 2, for a 10 ms frame, the transport latency will be 10 ms, plus the CIG Sync Delay. The example given in Figure 7.1 is with FT equals to one, so the transport latency in this example is equal to the CIG Sync Delay.

CIS may be configured to use a PHY type different from ACL, so, for example, in Figure 7.1, the ACL connection may be using a Bluetooth LE PHY type of 1 MHz bandwidth, while CIS may be using a PHY type of 2 MHz bandwidth. When using a 2 MHz PHY type for music streaming, the duty cycle of airtime usage is lower. This is essential for music streaming when PDUs are relatively longer. So, in the example in Figure 7.1, if the ISO interval is 10 ms and the PHY type is 2 MHz, the ideal airtime usage when no retries are used is ~16% when the bitrate is 192 kbps in total. That means that sending a CIG event with left and right PDUs to Peripherals 1 and 2 and receiving an acknowledgment takes around ~1600 µs out of a CIS interval of 10,000 µs. In practice, a higher duty cycle is used when PDUs are retried. In general, the Bluetooth LE link layer strives to keep the packet error rate low by applying link quality measures such as AFH and Power control. The Bluetooth® SIG is also working on future enhancements that will allow modifying the CIS parameters as a measure of link quality while streaming.

BIS LOGICAL TRANSPORT

The BIS logical transport is set up between one Isochronous Broadcaster and many Synchronized Receivers. When the application configures a device as an LE Audio BAP Broadcast Source, it configures the link layer to create a BIG. BIG contains a set of Audio Streams. BIG may contain one or more BISes (Broadcast Isochronous Streams).

The BIS logical transport carries any data that the application may send according to the QoS parameters, which were set for latency, bitrate, and reliability. The same methods that are used for ISOAL over CIS are also used for ISOAL over BIS. In the case of BIS, ISOAL is used to send data from the Isochronous Broadcaster side and receive data at the Synchronized Receiver. The Synchronized Receiver is not sending any Audio Stream link layer data back to the Isochronous Broadcaster. There are other mechanisms, however, in which a remote device may connect to a Synchronized Receiver in order to configure it over an ACL connection, so it may sync to the Isochronous Broadcaster. In certain use cases, the remote device may be the BIG Isochronous Broadcaster, while in other use cases, it may be another personal device that acts as a remote control application.

An example of a BIS Isochronous Broadcaster is a quiet TV in a public location. The TV will be using BAP to configure a BASE with the program info metadata, Codec configuration, and QoS configuration. The TV will configure the BASE information as Host Advertising data and send it in PA link layer messages. The TV will then use BAP to create BIG QoS parameters of latency, bitrate, and reliability to match the audio timing and Codec settings in the advertised BASE. While creating BIG, the PA is updated by the controller to include the BIGInfo element in the ACAD part of the PA. The BIGInfo provides the timing and scheduling parameters of the BIG with one or more BISes.

Figure 7.2 shows an example of two BIS logical transports used by an Isochronous Broadcaster device when broadcasting audio from a TV. In this example, the TV Audio Broadcaster is carrying different content, such as left and right stereo music streams or two different languages. Any listener devices may synchronize to the Broadcast audio. In the case that the two BIS streams carry left and right streams, then different people with left and right earbuds may synchronize the stereo TV streams. Alternatively, if the two BISes carry different languages encoded as two mono streams, then each person may synchronize to the BIS stream with the language of choice (e.g., English or Spanish). In the case of stereo music, a single device such as a headset with two speakers may synchronize to receive both BIS-1 and BIS-2 to play back the stereo content in two different speakers. The advertised BASE contains information about which BIS is carrying what audio channel, what encoding method, and what context, such as what TV program and language code. The PA also contains link layer information about how to synchronize to the various BIS indices; this link layer information is called BIGInfo.

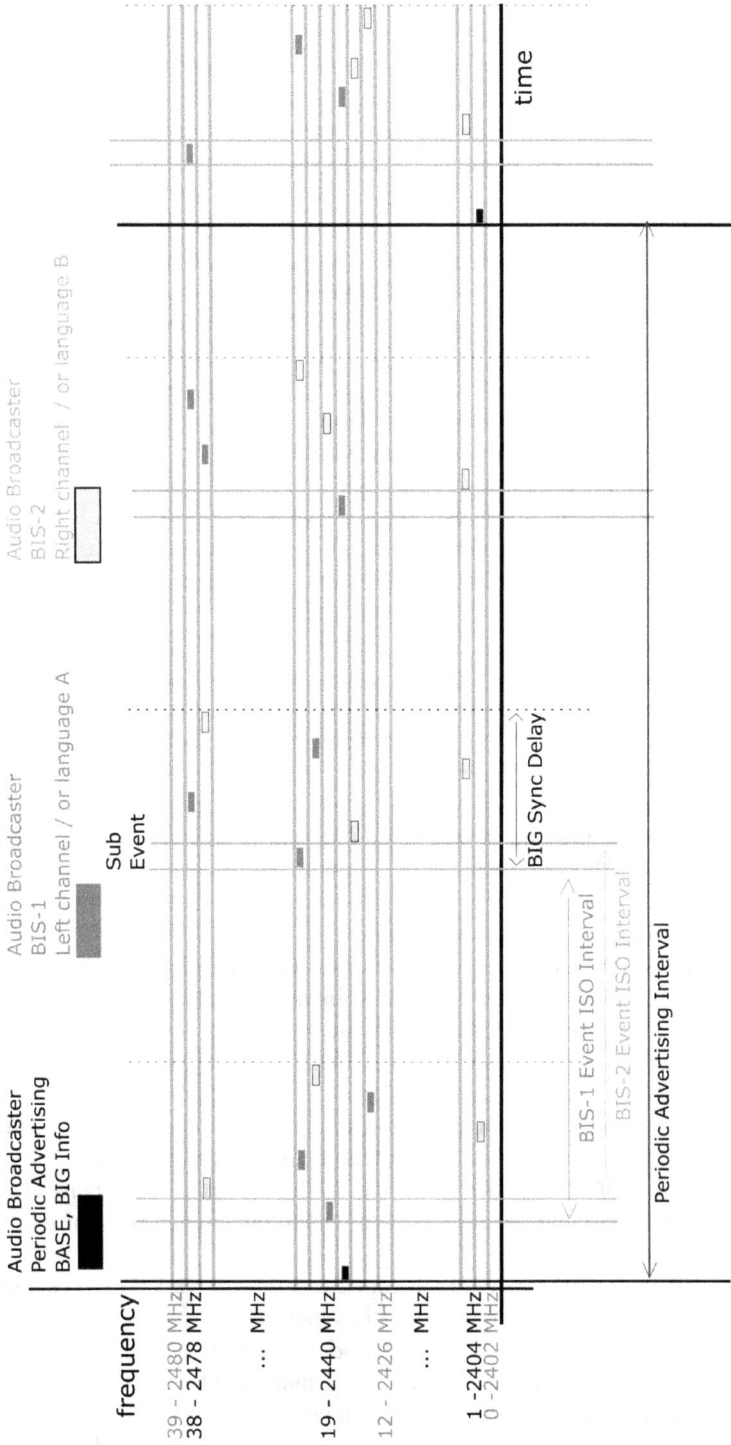

FIGURE 7.2 BIS Events over time and frequency.

The example given in Figure 7.2 shows that each BIS interval to BIS-1 or BIS-2 consists of different subevents. The Isochronous Broadcaster alternates the subevents to send BIS-1 and BIS-2 PDUs in an interleaved fashion. Each subevent is an unconditional retry of the same content of either BIS-1 or BIS-2. In Broadcast audio, there is no feedback mechanism because the content is sent to an unlimited number of listeners. In order to ensure reliability, the Isochronous Broadcaster sends the same content multiple times as a fixed Immediate Retransmission Count (IRC). In this example, in each BIS interval, the content is sent three times on each BIS, so IRC is set to three, and each of the subevents in each BIS contains a copy of the same content. The default inter frame space between the consecutive TX BIS PDUs within each subevent is 150 µs. Subevents may also contain start and continuation BN of audio data on top of IRC. The example in Figure 7.2 focuses on BN = 1 and IRC = 3, so all three numbers of subevents (NSE) are used for retries. If a listener receives the first transmissions with no checksum errors, then it may stop listening until the next BIS interval. However, the Isochronous Broadcaster does not know who received the content correctly, so it continues to retry the content two more times. So if, for example, other listeners receive the first and second content as corrupted packets due to local noise, then these receivers still listen for the third transmission. This mechanism adds reliability to the Broadcasted audio content. Note, however, that the Isochronous Broadcaster invests more power and uses higher bandwidth since it needs to send unconditional retries three times for every new encoded content. Unlike CIS, in BIS, a higher duty cycle is required. In this example, the duty cycle is around 40% out of the available air bandwidth. There is no bandwidth required for return packets, so the required bandwidth is not three times compared to CIS (~16% when no retries used), but the transmissions from the Isochronous Broadcaster contain long and repeatedly played audio content.

For high-reliability music applications, which may allow longer transport latency, the link layer provides a mechanism for a pre-transmission offset (PTO). If PTO is greater than zero, then the Isochronous Broadcaster may assign additional subevents in order to play a buffered source content ahead of time. So BIS event N may contain subevents with audio data intended for BIS event N+M, where, for example, if PTO is one, then M is one, and each BIS event also contains a pre-transmitted audio intended for the next interval. This mechanism adds time diversity, so it allows receivers an opportunity to receive audio content earlier, by a few intervals, than the rendering time. The Peripherals may buffer the early audio data for a later interval, so it adds more receive opportunities. If the pre-transmitted audio is received correctly, it also allows the receiver to not listen at the beginning of the next interval and instead try to get the next pre-transmitted audio in this event. The example in Figure 7.2 focuses on PTO equals zero, in which no pre-transmissions are done and in which low latency is required for an application such as a TV set.

BIS-1 and BIS-2 in Figure 7.2 are grouped together into a BIG. The Group of streams ensures that time-bound content is synchronized by the transport. For example, left and right stereo channels are sent as a group of BIS-1 and BIS-2, so the sampled content arrives as a set of earbud peripherals together. Upper profiles communicate a shared presentation delay so that the two peripherals also render the audio at the same time. The BIG Sync delay takes into account the entire transmissions to both BIS-1 and BIS-2. All Synchronized Receivers are aware of the BIG Sync delay and will render the audio based on this transport reference point.

The tool that the upper profile uses to communicate the information about the streams, the presentation delay, and the context and QoS information is Periodic Advertisements. Control information about the Isochronous Broadcaster stream is advertised in Periodic Advertisement packets. The Synchronized Receiver scans and synchronizes to the Periodic Advertisement, to receive the BASE property and the BIGInfo property, which allows it to get information such as details about the TV program, the Codec settings, and the timing of the BIG. This is shown in Figure 7.2 as the TV Periodic Advertisement interval. By Advertising the Broadcast audio in fixed intervals, any new listeners may sync and tune in to listen to the same Broadcast stream.

Table 7.1 summarizes the Periodic Advertisement content for Broadcast audio. It shows the BIGInfo and BASE properties and their content. BIGInfo is part of the Controller domain, and sent as ACAD within the PA payload. BASE is part of the Host profile and sent as Host Advertising data within the PA payload.

TABLE 7.1
Periodic Advertisement Content

Property	Domain/Advertising Part	Purpose
BIG Info	Controller Additional Controller Advertising Data	Information about the timing of the BIG, Access Address of the BISes, the interval of the BIG, and timing of each BIS in the BIG. Information about IRC, PTO, NSE, BIG Sync Delay, maximum size of each packet, PHY type, channel mask, and optional public key for encryption
BASE	Host profile Host Advertising data	Information about Codec settings, presentation delay, frame sizes, bitrates, audio locations of each BIS in a BIG, content of each BIS in a BIG, context type, and metadata about the stream, such as TV Program Info

CIS CONTROL

Figure 7.3 shows the creation time of the CIG and CIS. The link layer uses a three-way handshake mechanism to create each CIS in a CIG, using the following link layer control messages over the ACL transport to each Peripheral (ACL-1 and ACL-2):

- CIS Request
- CIS Response
- CIS Indication

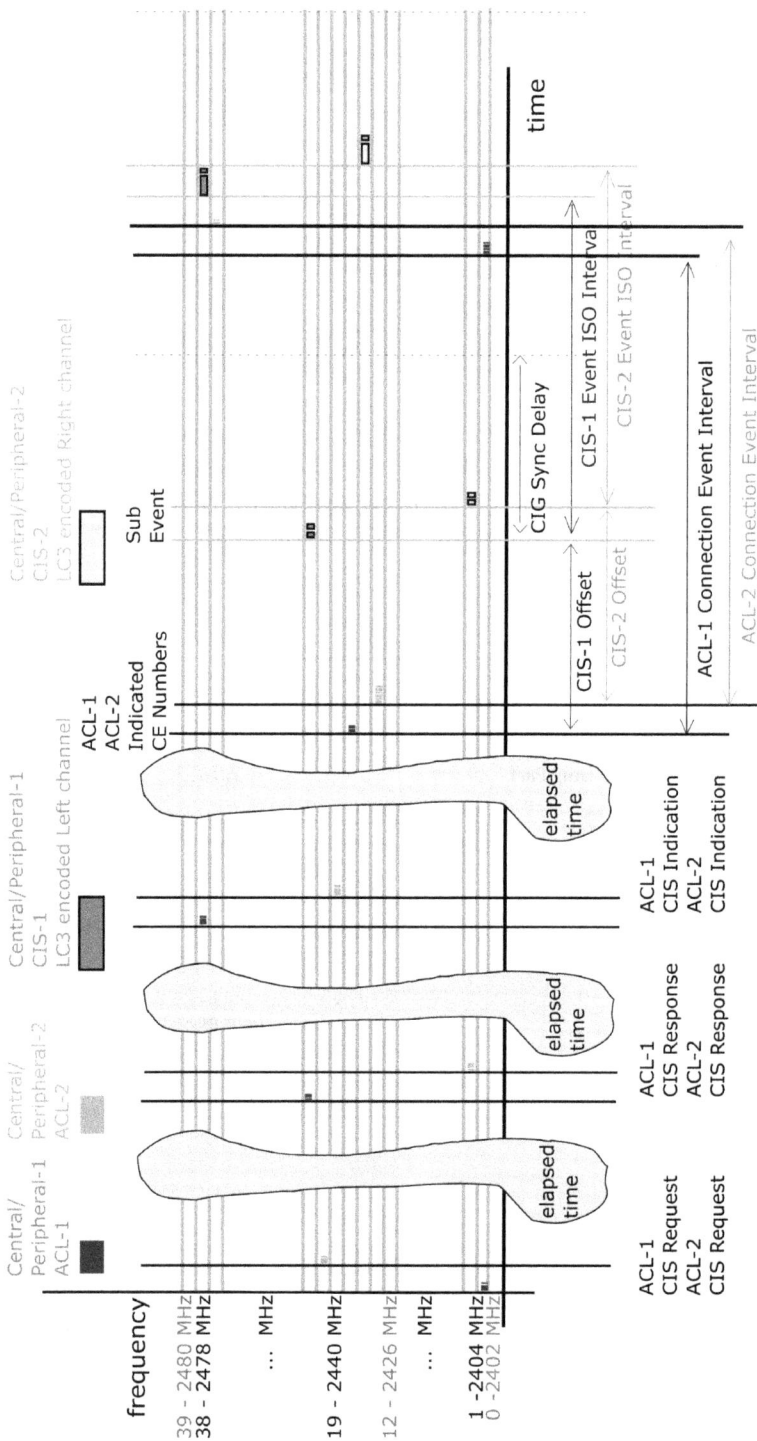

FIGURE 7.3 CIG/CIS Creation.

The CIS Request is sent by the Central to each of the Peripherals. The CIS Request contains all the information and parameters about the CIS, such as NSE, BN, FT, ISO interval, SDU, and PDU max sizes, and it also proposes an instant time in the future where the first CIS anchor begins. The time is given as a CIS Offset from a future ACL Connection Event (CE) Number. The Central sends a CIS Request to each Peripheral as part of the CIG. The parameters about the number of CIS in a CIG and which Peripherals belong to each CIS are configured by the Host BAP profile. The Host also requests the QoS parameters (bitrate, reliability, latency, SDU size, SDU interval, PHY type), which the Controller translates to scheduling parameters (NSE, BN, FT, packet size, interval, PHY type). When the BAP profile needs to begin streaming, it commands the Controller to create CISes to each Peripheral in a CIG, which forms an audio group of devices.

The CIS Response is sent back from each Peripheral after the Peripheral Host accepts the new CIS. When the Peripheral receives the CIS Request from the Central, it sends an event to the BAP Server and requests approval of the new CIS; BAP responds back to approve the CIS, and then the Peripheral sends the CIS Response link layer control message. Figure 7.3 illustrates that there is an elapsed time in sending a CIS Request from the Central to the Peripheral until the Peripheral responds back with a CIS Response. In the CIS Response, the Peripheral provides a proposed anchor point in the form of an ACL CE Number and CIS Offset valid range for the Peripheral.

When the Central receives the CIS Response, which approves the new CIS, then the Central checks if it needs to recalculate a new proper offset, considering the proposal from the Peripheral in the CIS response and any link layer delays since the previous sent CIS Request. It could happen that the previous instant is in the past due to link layer retries or a long delay to accept the CIS. The Central then sends a CIS Indication with the final CIS anchor point as a CIS Offset from a future ACL CE Number. In this message, the Central also assigns a new Access Address for the CIS. The CIS Access Address is the first part of Bluetooth® LE PDU, and it is different from the ACL Access Address to the same Peripheral, and also different from the Access Address of other CISes to a different Peripheral. Each Peripheral syncs on the Access Address, which assures that content is arriving at the intended Peripheral on the intended transport (CIS or ACL). The Access Address is a 32-bit random number. And all CIS PDUs will contain the same Access Address, which is assigned in the CIS Indication. The Access Address is also used to derive unique RF frequencies for each subevent in a CIS event based on the Bluetooth® LE Channel Selection Algorithm #2 from Bluetooth Core 5.

The three-way handshake described earlier is done separately with each Peripheral. Once the final CIS Indication is acknowledged by each Peripheral, then all parties know the exact first anchor point to begin streaming audio data over CIS-1 and CIS-2, as an offset to the Bluetooth LE ACL-1 and Bluetooth LE ACL-2 transports. Apart from using the ACL connection to establish CIS to each participating Peripheral, the ACL connection also serves the upper profiles for service discovery and service setup. When the use case is idle, in standby, and no audio is playing, the ACL connection may disconnect or may remain connected in low-power mode. When in low-power mode, the Central or Peripheral may initiate a procedure to increase the ACL Connection Event interval to save more power. The Bluetooth SIG enabled a new Core feature called ECU. ECU allows rapid changes between low-power mode (longer ACL intervals) and low-latency mode (shorter ACL intervals). Switching to low-latency mode is required when a use case needs to begin, or a service access is required by the upper profile. Other essential features which were added by the Bluetooth SIG are Bluetooth LE Power Control and Bluetooth LE AFH Peripheral classification. In audio use cases, it is likely that a user who is wearing a headset or earbuds will walk away from a phone or PC, or a user with a phone or PC will walk away from a wireless speaker. With Bluetooth LE Power Control, the power levels used by the radio are optimized such that low power is used when devices are closer, and higher power is used when devices are further apart. The Bluetooth LE Power Control helps to maintain a robust connection while saving power in both Central and Peripheral. The Bluetooth LE AFH channel classification allows for Peripherals to classify bad RF channels, so the Central can take the feedback from each Peripheral into account when defining a new AFH channel map for each Peripheral. This feature is essential to preserve adequate link quality, since, as we saw earlier, in LE Audio the topology may involve multiple speakers in various locations, or a user who is wearing a headset, and is further away from a phone, TV, or a PC. In these cases, the interference levels that the Central observes are different from the interference levels that the Peripherals observe (as Peripherals reside in a different location). With Bluetooth LE AFH channel classification, the Central is able to accommodate good RF channel selection by taking into account feedback from all Peripherals in a use case (one or more). The Central will be able to define a dedicated AFH channel map toward each Peripheral per each ACL connection, so each connection to a Peripheral preserves good link quality conditions. To summarize, the ACL connection between a Central and each Peripheral serves a critical role in the profile signaling handshake during service discovery, endpoint discovery, and service setup to begin the audio use case (ASE establishment). The ACL connection is enhanced to allow quick interval changes to save idle power, power control to optimize power vs. distance, and AFH classification per each peripheral to assure the RF radio link quality is adequate.

BIS CONTROL

Figure 7.4 shows an example of how a Synchronized Receiver may synchronize to a BIG and begin receiving Broadcast audio data. The synchronization process consists of the following steps:

- Scanning for primary Advertising
- Scanning for secondary Advertising, detecting a BAP Broadcast Source UUID, also known as Broadcast Announcement in BAP (see Chapter 5)
- Syncing to Periodic Advertising, retrieving BIGInfo, BASE. The BASE is also known as Basic Announcement in BAP (see Chapter 5)
- Syncing to the BIG

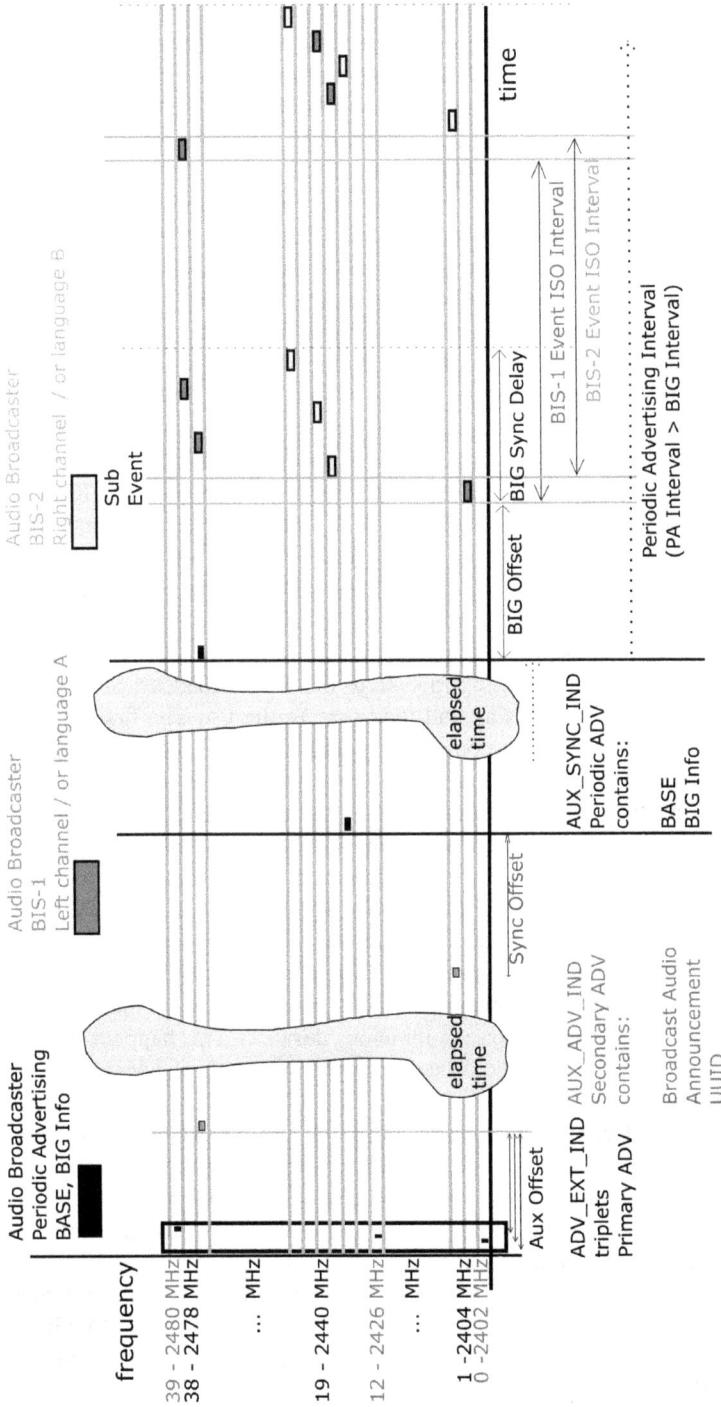

FIGURE 7.4 Synchronizing to a BIG.

Each of the preceding steps requires a turnaround time between the Host profile and the Controller. The first two steps of scanning for primary and secondary Advertising take the most amount of time, because the scanner needs to sort through many Advertising reports until it finds a secondary Advertising that contains a BAP Broadcast Source UUID. Only then are these reports processed by the Host profile, which then commands the Controller to sync to the Periodic Advertising, which is pointed by the secondary Advertising by the Sync Offset. Then once again, when the Controller syncs to the PA, it sends an Advertising report about the Periodic Advertisement and a BIGInfo report about the BIG. The Host profile makes a decision to sync to this BAP Broadcast Source based on the BASE information. Once it decides to sync, it commands the Controller to sync to the BIG.

The BIG logical transport is connectionless. The BAP Broadcast Source is Advertising about BIG in order for remote listeners to sync to one or more BISes in a BIG. The BAP Broadcast Source uses the primary Advertising channels (triplets) to point to an extended secondary Advertising channel, via an AUX Offset. The extended Advertising channel contains host data that announces the presence of a Broadcast. The Announcement is a UUID assigned by the Bluetooth® SIG to identify the Broadcast Audio Announcement. This allows remote devices to identify that this extended Advertising is originating from a BAP Broadcast Source. The extended Advertising is further pointing to a PA via a Sync Offset. The extended Advertising may also contain a friendly name which the Host can use to present in a user interface for a user to select.

A remote device that is looking to sync to the BAP Broadcast Source will first identify the extended Advertising and then sync to the PA. The first step requires a synchronizing device to scan continuously until it finds a secondary Advertising that contains a BAP Broadcast Source. The first two steps of scanning for primary Advertising and then for secondary Advertising may consume large amounts of power from small earbuds or hearing aid devices. Therefore, there is an additional mechanism that allows a third device to scan for BAP Broadcast Sources and pass the information about the PA to the earbuds or hearing aids. This mechanism is known as the PAST. The third device is known as the scan assistant (or Broadcast assistant) and connects over ACL to the BAP Broadcast Sink and passes the information over the link layer. After this procedure is done, the BAP Broadcast Sink may sync to the PA without having to scan for a long duration. This happens by receiving offset information to the Periodic Advertisement from the third device over the ACL logical transport.

The Periodic Advertisement contains both link layer information about the BIG/BIS logical transport (BIGInfo) and host information about the audio (BASE). The link layer information contains timing of where the BIS is transmitted in relation to the Periodic Advertisement event, and information about size and QoS info, such as retry info and payload size info. The host information in the Periodic Advertisement contains information about the Codec settings, location, context, and metadata from the application. When a remote device is synchronizing to the BAP Broadcast Source Periodic Advertisement, it has all the information to sync to one or more BISes in a BIG.

Once synchronized, the remote device (the BAP Broadcast Sink) continues to receive the BIS logical transport, as a Synchronized Receiver, in regular intervals, which are spaced equally apart from each other. In case the remote device lost sync or went through a power cycle, it may follow the synchronization procedure again by scanning for the Periodic Advertisement to regain the BIS's timing.

The Synchronized Receiver may choose to remain synchronized to the Periodic Advertisement when BIG stops streaming, so it may be ready to receive Broadcast audio again when streaming is resumed and BIGInfo is added to the Periodic Advertisement ACAD. The presence or absence of BIGInfo indicates when BIG is present. In addition, the BIG contains dedicated control subevents to indicate the update of the channel map or termination of the BIG. However, when BIS is removed, the only way to tell when it is reestablished again at a later time is to either stay synced to the Periodic Advertisement or go back to scanning.

REFERENCE

1. Bluetooth® Core 6.0 or later, https://www.bluetooth.com/specifications/specs/core-specification-6-0/

8 Vision

BRING IT ALL TOGETHER

As we saw in the previous chapters, the LE Audio One Architecture consists of dedicated building blocks to address and serve specific purposes. There are transport blocks, Codec blocks, and building blocks for the discovery of Codec capabilities and the discovery of endpoints. There are other building blocks to configure audio endpoints. There are building blocks to coordinate multiple devices, control their volume, and allow remote control of the media player and voice gateways.

The LE Audio One Architecture defines how to commonly use the controls by a use case-based profile. The building blocks allow a use case-based profile to access audio and then optimize audio to an existing use case or add a profile to support future use cases. The LE Audio One Architecture is providing this toolbox of LEGO blocks and the procedures on how to connect the blocks. The LE Audio One Architecture does not limit the use cases to certain combinations. A few use case-based profiles are provided, and it is expected that a new set of use case-based profiles will emerge in the coming years.

At this point, it is good to go back to the high-level partition of the LE Audio One Architecture from Chapter 1. In Chapter 1, we described the LE Audio Architecture as it evolved from Classic Audio into LE Audio. We also defined the architecture layers: App, Control, and Transport. Later in the book, we went into more detail about each layer and each block.

Figure 8.1 shows the resulting high-level LE Audio Architecture with the actual block names, which we described in the previous chapters.

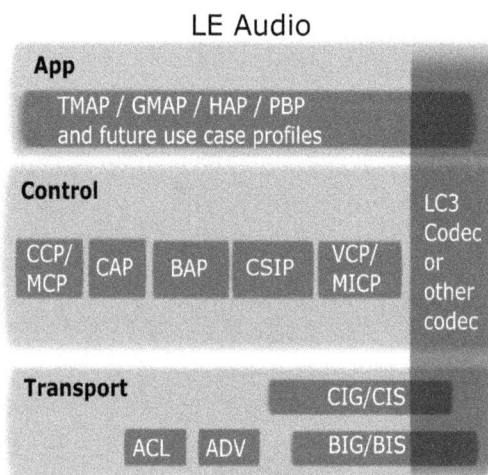

FIGURE 8.1 LE Audio One Architecture with block names.

DOI: 10.1201/9781003590187-8

The App layer consists of TMAP/GMAP/HAP/PBP, while future use case-based profiles are expected. TMAP is the high-quality audio profile for media playback and call gateway use cases supporting both Unicast audio and Broadcast audio. GMAP enables a profile to mix both media audio and voice audio over the same stream, enabling applications such as gaming in low-latency and high-quality audio. HAP is a profile targeting hearing aids. PBP is addressing public broadcast use cases, as well as encrypted personal audio share, and it is the foundation for Auracast™ broadcast audio in public locations.

The Control layer provides basic building blocks to generically activate and start audio. CCP and MCP are the call control and media control. CAP defines procedures on a set of devices and how to use the other controls to achieve a general use case. BAP is controlling Audio Streams. CSIP coordinates bonding to a set of audio peripherals. VCP and MICP control the volume and gain of speakers and microphones.

The Transport layer provides over-the-air streams, which are CIG/CIS for Unicast audio and BIG/BIS for Broadcast audio, and the connectivity transport for profile-level signaling and Advertising an announced Audio Context.

LC3 or other codecs are used across the architecture layers to carry compressed audio from an application to or from the antenna. Every layer is interacting with different aspects of the codecs: audio content (App layer), audio quality (App layer), configuration of the Codec settings (Control layer), QoS to the Codec settings (Transport layer).

FUTURE USE CASE EXAMPLES

As seen in Figure 8.1, the LE Audio One Architecture provides the App, Control, Codec, and Transport, which allows defining new use case-based profiles. The building blocks are generic and are used by the use case-based profiles via common procedures, which are specified in CAP. The common procedures may be used to derive new use cases. In the next few sections, we will provide examples for future use cases that may evolve over LE Audio. This is a thought exercise, in which a given use case is described in terms of required behavior. The LE Audio architectural layers make the following use cases possible.

Museum

Imagine yourself walking inside a museum, which shows a new exhibition that you were expecting to see for quite some time. In the ticket booth, you get the name of the Broadcast guide. It's called "Leonardo's greatest inventions." As you walk inside the lobby, you are wearing your LE Audio earbuds, one in each ear. Soon after, you hear a voice, coming from your earbuds, which lets you know about an LE Audio Broadcast source found by your mobile phone. You take your phone out of your pocket and open the alert application.

Your phone shows you a list of LE Audio Broadcast sources. You scroll through a few exhibition tour guide options until you notice the line: "Leonardo's greatest inventions." Aha, you select this option, and a voice in your earbuds is letting you know that the tour begins in five minutes, at exhibition hall L. You step toward exhibition hall L while a background mandolin medieval music is playing in your earbuds, and you lower the volume to a level that allows background listening. Once you arrive at the entrance of exhibition hall L, a voice is letting you know that the exhibition begins in one minute.

You enter the hall and begin listening to a biography guide about Leonardo's early childhood, while you increase the volume and look at some instruments which are presented inside hall L. You continue to look at some of Leonardo's rare childhood paintings when the Broadcast guide reminds you that in exhibition hall E, a live show is about to begin, and the topic is: Leonardo at the guild. You start walking toward exhibition hall E while a mandolin theme is playing in your earbuds.

Table 8.1 describes how the museum use case is realized using the LE Audio One Architecture.

TABLE 8.1
Museum Use Case Layers, Components, and Roles

Layer	Module	Role	Description
App	TMAP	BMS	The museum deploys a multimedia public announcement system to broadcast high-quality audio over Bluetooth®.
		BMR	The visitors of the museum use their earbuds or headsets to synchronize to the Broadcast.
	PBP	PBS	The museum deploys a dedicated hearing aid public announcement system, with a Public Broadcast Source transmitting at standard quality.
	HAP	HA	Visitors with hearing loss use their hearing aids to synchronize to the hearing aid announcements.
		HARC	Visitors with hearing aids use their cellphones to select the dedicated hearing aid public announcements.
Control	BAP	Broadcast Source	The museum sets up a high-quality multimedia Broadcast stream and a lower-quality multimedia Broadcast stream for hearing aid users; the BASE structure is advertised with Codec settings and metadata settings such as the name of the tour guide.
		Broadcast Sink	The museum visitors synchronize their audio earbuds or hearing aids to a Broadcast stream they select from a list after delegate scanning to a Broadcast assistant application on a cellphone.
		Broadcast Assistant	The museum visitors select the Broadcast source via an application on a cellphone to apply on both earbuds as managed by CAP.

(Continued)

TABLE 8.1 *(Continued)*
Museum Use Case Layers, Components, and Roles

Layer	Module	Role	Description
	VCP	VCP Controller	The museum visitors use their cellphones to change the volume of their earbuds, hearing aids, or headsets.
		VCP Renderer	The headsets or earbuds apply the volume settings for each user's liking.
	CSIP	Set Coordinator	Museum visitors with a set of earbuds or a set of hearing aids use their cellphones to connect to their authenticated set.
		Set members	The pair of earbuds or the pair of hearing aids is authenticated for each visitor.
	CAP	Initiator	The museum creates multiple streams per tour guide type and audio quality.
		Acceptor	The users synchronize to a Broadcast stream with matching left and right stereo properties.
		Commander	Visitors use an application on their cellphone to connect to a set of earbuds or hearing aids, add a Broadcast source, and apply matching BAP stream settings and matching volume settings to the set of surround speakers.
Codec	LC3	High-quality media settings	The museum deploys the multimedia tour guide as an LC3 high-quality with a full band sampling rate of 48 kHz.
		Hearing aid media settings	The museum deploys the hearing aid version of the tour guide as LC3 mid-quality with a semi-super wideband sampling rate of 24 kHz, which enables people with hearing loss to consume all audio content at lower tones.
Transport	BIG/BIS	Broadcaster	The museum announcement system sends audio over Broadcast Isochronous Streams and publishes the content and settings over a Periodic Advertising carrying controller timing in BIGInfo and BASE structure with Codec settings and metadata settings such as the name of the tour guide.
		Synchronized Receiver	The visitors synchronize their earbuds or hearing aids to the Broadcast Isochronous stream.
		Central	The users use their phone to connect, read PAC records for Codec capabilities, and adjust volume settings of the earbuds, hearing aids, or headset; based on PAC records, the phone selects the Broadcast source, configures the BASS server to synchronize to the source, and offloads scanning from peripherals via a PAST controller link layer procedure.
		Peripheral	The earbuds, hearing aids, or headsets publish PAC records to the phone, receive volume settings from the phone, and delegate scanning via a BAP BASS service and Controller link layer PAST procedure.

HOME THEATER

A few days later, one evening, you walk into a living room with your PC in your hand. You are planning to watch a new movie with a friend. Your friend told you that she has a new set of LE Audio surround speakers, and you suggested giving them a try with your LE Audio-enabled PC. The set includes a sound bar piece, a subwoofer piece, a rear left piece, and a rear right piece. You and your friend place the speakers around the living room, press a button on each piece, and start the PC discovery of the speaker set. You begin playing the movie on the PC while projecting the picture on the living room wall and immerse yourself in multichannel surround LE Audio.

Table 8.2 describes how the home theater use case is realized using the LE Audio One Architecture.

TABLE 8.2

Home Theater Use Case Layers, Components, and Roles

Layer	Module	Role	Description
App	TMAP	BMS	The PC deploys multichannel surround and applies multichannel calibration settings, balancing, and leveling of volume across the surround configuration.
		BMR	Each speaker is controlled by the PC to provide the best volume setting, given its properties and reported configuration.
Control	BAP	Broadcast Source	The PC sets up a single Broadcast stream to carry the multichannel surround signal.
		Broadcast Sink	The speakers synchronize to a Broadcast stream as instructed by the PC; each speaker retrieves the location-specific signals from the multichannel Broadcast stream.
		Broadcast Assistant	The PC selects the Broadcast source to be itself and configures each speaker to sync to the Broadcast.
	VCP	VCP Controller	The PC controls each speaker volume per the surround configuration.
		VCP Renderer	The speaker volume is adjusted per optimal calibrated configuration.
	CSIP	Set Coordinator	The PC discovers the set of four speakers, authenticates to them as a set, connects to each one of the speakers, and enables data encryption.
		Set members	The four speakers publish themselves as part of a single set.
	CAP	Initiator	The PC configures to start audio as Broadcast.
		Acceptor	The speakers synchronize to a Broadcast stream.
		Commander	The PC connects to a set of speakers, adds a Broadcast source as itself (the PC), adds the Broadcast Code for encrypted personal audio, and applies matching BAP stream settings and matching volume settings to the set of surround speakers.

(Continued)

TABLE 8.2 *(Continued)*

Home Theater Use Case Layers, Components, and Roles

Layer	Module	Role	Description
Codec	Multichannel Codec (vendor-specific)	Surround settings	The PC deploys the multichannel encoded content such that each speaker may synchronize to the same Broadcasted stream while extracting the relevant location to render (subwoofer, rear left, rear right, and a sound bar with front left, front right, and center).
Transport	BIG/BIS	Broadcaster	The PC sends audio over a single Broadcast Isochronous Stream.
		Synchronized Receiver	The speakers synchronize to the Broadcast Isochronous stream.
		Central	The PC connects and controls all speakers and offloads scanning from speakers via PAST to allow them to synchronize to the PC Broadcast stream. Other settings, such as profile data from TMAP and VCP, are sent to each speaker.
		Peripheral	The speakers receive configuration from the PC and delegate scanning via PAST.

Online Music Jam

After the movie is over, the two of you decide to play music together with friends you know from other countries. You start a video conference with two friends in two different countries while connecting to the speaker set. You turn on two stereo microphones on the rear left and rear right speakers and on the sound bar. When your online friends join the video conference, you can see them projected on the living room while hearing them in the multichannel surround speaker set. Each one of you picks up a favorite instrument, and you begin a live jam streaming concert. Your friend chose to be the singer and connected an additional LE Audio microphone to the rear left speaker for a better sound capture, via a button on each device. You walk around with the guitar strapped on your shoulder and make sure to stay between the rear left and rear right speakers, where the two stereo microphones are located.

Table 8.3 describes how the online music jam use case is realized using the LE Audio One Architecture.

TABLE 8.3

Online Music Jam Use Case Layers, Components, and Roles

Layer	Module	Role	Description
App	TMAP	UMS	The PC deploys a multichannel surround and applies multichannel calibration settings, balancing, and leveling of volume across the surround configuration while enabling microphones with low-latency settings.

(Continued)

TABLE 8.3 *(Continued)*
Online Music Jam Use Case Layers, Components, and Roles

Layer	Module	Role	Description
		UMR	Each speaker is controlled by the PC to provide the best volume setting, given its properties and reported configuration; two speakers enable microphones to capture audio from the room and from the additional singer wireless microphone.
Control	BAP	Unicast Client	The PC sets up four sink Unicast streams to carry the multichannel surround signal to the speakers with low latency, and two source Unicast streams to capture audio by the two speakers with the attached microphones, with low latency, and from the extension wireless microphone.
		Unicast Server	The four speakers publish their Codec capabilities and follow the PC to move their sink endpoints to a streaming state; two of the speakers also move source endpoints to a streaming state.
	VCP	VCP Controller	The PC controls each speaker volume per the surround configuration.
		VCP Renderer	The speaker volume is adjusted per optimal calibrated configuration.
	CSIP	Set Coordinator	The PC discovers the set of four speakers, authenticates them as a set, and connects to each one of the speakers to set encryption.
		Set members	The four speakers are published as part of a single set.
	CAP	Initiator	The PC is configured to start audio as Unicast with the same sink settings on all four speakers while configuring their audio locations accordingly; source settings are configured to capture high-quality audio from the room singer and guitarist.
		Acceptor	The speakers accept each Unicast stream setting and Unicast source setting.
Codec	LC3	Source/Sink	The PC deploys four sink settings to 48 kHz and two source settings to 48 kHz to render and capture at highest LC3 quality and at low latency.
Transport	CIG/CIS	CIS Central	The PC forms a CIG and sends audio over Connected Isochronous Streams to all four speakers while receiving audio from two speakers and the extension microphone.
		CIS Peripheral	All four speakers receive audio over separate CISes; two of the speakers also send audio back to the PC over their bidirectionally configured CISes while forwarding mixed captured audio via the extension wireless microphone proxy.
		ACL Central	The PC connects and controls settings such as profile data from TMAP, BAP, CSIP, and VCP, which are sent to each speaker.
		ACL Peripheral	The speakers receive configuration from the PC.

Interactive Classroom

A few weeks later, you join a frontal course at the local community center. The course is on the history of the human race and begins with a series of lectures on the ice age. In one of the lessons, the lecturer partitions the class into two parts. The rear part of the class is watching a documentary movie about the ice age, while the front part of the class is focusing on a guided assignment. The lecturer needs to help the students in the guided assignment, so a smaller part of the class allows the lecturer to help each student, one at a time. The rear part of the class is using their LE Audio headsets to tune to the TV, which is showing the documentary about the ice age. The front part of the class is not seeing the TV, nor hearing it, thanks to the Broadcasted audio over LE Audio. The lecturer is happy to use the time efficiently while allowing the students who are doing an assignment to focus quietly, while the other students watch and listen to the documentary. This allows the lecturer more time to help the smaller group of students with their assignment. Later, the students switch places, so each part can see the movie and do the guided assignment with the lecturer. In the last hour, the entire class is conducting a lively discussion about the movie and the assignments. You are impressed by how efficient a class can be. In a single lesson, you heard a lecture, watched a short documentary, completed an assignment with the lecturer's help, and participated in a lively discussion.

Table 8.4 describes how the interactive classroom use case is realized using the LE Audio One Architecture.

TABLE 8.4

Interactive Classroom Use Case Layers, Components, and Roles

Layer	Module	Role	Description
App	PBP	PBS	The TV is configured for public Broadcast and publishes program info metadata about the "Ice Age" class title. The TV announces two audio quality levels: high quality 48 kHz, and standard quality 24 kHz (for hearing aid users).
		PBK	Each one of the students wears a headset that synchronizes to the Broadcast TV stream.
		PBA	The students use their cellphones to find the classroom TV Broadcast title: "Ice Age."
Control	BAP	Broadcast Source	The TV sets up a Broadcast stream to carry the TV audio content and an additional stream for hearing aid users and advertises the content and settings over a Periodic Advertising carrying controller timing in BIGInfo and BASE structure with Codec settings and metadata program info settings such as the name of the program "Ice Age."
		Broadcast Sink	The headset synchronizes to a Broadcast stream as instructed by each student's phone. In case there are students with hearing loss, they tune their hearing aids to the hearing aid TV stream.
		Broadcast Assistant	The cellphone selects the TV Broadcast stream titled "Ice Age" or "Ice Age (hearing aid quality)" in the case of a student with hearing loss.
	VCP	VCP Controller	The student's phone controls their headset volume, while each student tunes the volume differently to their liking.
		VCP Renderer	The headset volume is adjusted.
	CAP	Initiator	The TV is configured to start audio as a TV high-quality context type and another stream with hearing aid quality.
		Acceptor	The headsets, earbuds, or hearing aids synchronize to a Broadcast stream.

(Continued)

TABLE 8.4 *(Continued)*
Interactive Classroom Use Case Layers, Components, and Roles

Layer	Module	Role	Description
		Commander	The phone commands the Acceptor to begin receiver synchronization to the TV and controls the headset volume.
Codec	LC3	media settings	The Broadcast stream is configured to a media setting of 48 kHz; a different stream at a lower sampling rate of 24 kHz is offered in case there are students with hearing loss who use hearing aids.
Transport	BIG/BIS	Broadcaster	The TV sends audio over Broadcast Isochronous Streams and advertises the content and settings over a Periodic Advertising carrying controller timing in BIGInfo and BASE structure with Codec settings and metadata program info settings, such as the name of the program "Ice Age."
		Synchronized Receiver	The students synchronize their headsets or hearing aids to the Broadcast Isochronous stream.
		Central	The users use their phone to connect and control the earbuds, hearing aids, or headsets, and assist scanning towards peripherals via PAST.
		Peripheral	The earbuds, hearing aids, or headsets receive configuration from the phone and delegate scanning via PAST.

TRADE SHOW

You attend an international world fair abroad. The fair is held in a large outdoor park in a major city. The fair has hundreds of booths. While you walk around, you realize that there are also special demonstrations that are happening at certain times. You would like to roam and watch as many booths as you can while also showing up on time for each demonstration. Luckily, you learn that the world fair installed an LE Audio public announcement system. You take your pair of earbuds out of your pocket and use your mobile phone to scan for the world fair public announcement system. Once your cellphone discovers the system, you select it and begin hearing a welcome English message coming from your earbuds.

You are happy to hear that the system also allows you to pick your favorite language out of 27 different languages. You use your mobile phone again to discover the stream that matches your native language. Once you select your language, you can continue strolling from booth to booth while getting voice notifications about upcoming demonstrations, their description, and location. You are happy that you can walk around, see booths, and talk to people around you while getting reminders to not miss the action of a new demonstration that you are expecting to see. Your earbuds guide you about each imminent demonstration in your native language. This world fair made you feel right at home.

Table 8.5 describes how the trade show use case is realized using the LE Audio One Architecture.

TABLE 8.5

Trade Show Use Case Layers, Components, and Roles

Layer	Module	Role	Description
App	TMAP	BMS	The trade show deploys a multimedia public announcement system to broadcast high-quality audio over Bluetooth® in multiple streams.
		BMR	The visitors of the trade show use their earbuds or headsets to synchronize to the Broadcast stream of their choice.
	PBP	PBS	The trade show deploys a dedicated hearing aid public announcement system, using standard quality at a 24 kHz sampling rate. Along with a high-quality system at 48 kHz.
	HAP	HA	Visitors with hearing loss use their hearing aids to synchronize to the hearing aid announcements.
		HARC	Visitors with hearing aids use their cellphones to select the dedicated hearing aid public announcements.
Control	BAP	Broadcast Source	The trade show sets up multiple high-quality multimedia Broadcast streams and multiple lower-quality multimedia Broadcast streams for hearing aid users; the BASE structure is advertised with Codec settings and metadata settings such as the name of the trade show, while each stream advertises a different BASE with the name in one of the 27 different languages via each language character set.
		Broadcast Sink	The trade show visitors synchronize their audio earbuds or hearing aids to a Broadcast stream they select from a list after delegate scanning to a Broadcast assistant application on a cellphone.
		Broadcast Assistant	The trade show visitors select the Broadcast source via an application on a cellphone to apply to the earbuds as managed by CAP; visitors select the language of choice based on the character set shown in the list of Broadcasters.
	VCP	VCP Controller	The trade show visitors use their cellphones to change the volume of their earbuds, hearing aids, or headsets.
		VCP Renderer	The headsets or earbuds apply the volume settings for each user's liking.
	CSIP	Set Coordinator	The trade show visitors with a set of earbuds or a set of hearing aids use their cellphones to connect to their authenticated set.
		Set members	The pair of earbuds or the pair of hearing aids is authenticated for each visitor.
	CAP	Initiator	The trade show creates multiple streams per language.
		Acceptor	The users synchronize to a Broadcast stream with matching left and right stereo properties.
		Commander	Visitors use an application on their cellphone to connect to a set of earbuds or hearing aids, add a Broadcast source, and apply matching BAP stream settings and matching volume settings to a set of earbuds or hearing aids.

(Continued)

TABLE 8.5 *(Continued)*
Trade Show Use Case Layers, Components, and Roles

Layer	Module	Role	Description
Codec	LC3	High-quality media settings	The trade show deploys the public announcement system as LC3 high quality with a full band sampling rate of 48 kHz.
		Hearing aid media settings	The trade show deploys the hearing aid version of the public announcement system as LC3 mid-quality with a semi-super wideband sampling rate of 24 kHz, which enables people with hearing loss to consume all audio content at lower tones.
Transport	BIG/BIS	Broadcaster	The trade show announcement system sends audio over Broadcast Isochronous Streams and publishes the content and settings over a Periodic Advertising carrying controller timing in BIGInfo and BASE structure with Codec settings and metadata settings with each language character set.
		Synchronized Receiver	The visitors synchronize their earbuds or hearing aids to the Broadcast Isochronous stream.
		Central	The users use their phones to connect and control the earbuds, hearing aids, or headsets, and offload scanning from peripherals via PAST.
		Peripheral	The earbuds, hearing aids, or headsets receive configuration from the phone and delegate scanning via PAST.

HOME RECORDING STUDIO

A few months later, you began a new hobby: a home recording studio. You invite friends and family to record a few songs. You just got a new LE Audio-enabled guitar. While you practice and rehearse toward recording, you learn that the guitar allows you to hook two additional audio inputs in order to allow a voice microphone and a harmonica microphone. It also has a USB to connect to an external synth. You start your mobile PC and install the home studio application. The application allows full control over the guitar input, the synth, and the mics from your voice or the harmonica. With a few clicks, you begin recording the new song while allowing different parts of the track to switch between the different audio inputs automatically. For example, during the guitar solo, the voice and harmonica microphones are turned off from the PC application, all wirelessly. In another section, the application balances input levels from the guitar, your voice, and the synth. You are playing the guitar and singing while a friend is tuning the application knobs, and a family member is hitting notes on the synth. All three of you are wearing LE Audio headsets, and when done recording, you start broadcasting the recorded song you composed from the PC to your headsets. You all listen together to the song you recorded just a few minutes ago.

Table 8.6 describes how the home recording studio use case is realized using the LE Audio One Architecture.

TABLE 8.6
Home Recording Studio Use Case Layers, Components, and Roles

Layer	Module	Role	Description
App	TMAP	UMS	The PC deploys multichannel topology across the recorded instruments and applies multichannel calibration settings, balancing, and leveling of volume across the channel configuration while enabling microphones with low-latency settings.
		UMR	Each instrument input is controlled by the PC to provide the best gain setting, given its properties and reported configuration.
Control	BAP	Unicast Client	The PC sets a Unicast audio connection to the guitar and configures multiple channels.
		Unicast Server	The guitar publishes its Codec capabilities and configures its source endpoint to a streaming state, each with multiple input locations.
	MCP	MCP Client	The guitar controls the recording studio application remotely while allowing the guitar player to switch tracks, start record, stop record, or begin playing the recorded track.
		MCP Server	The PC recording application switches tracks and starts/stops recording per the guitar user's choice or begins the playback of the recorded song.
	MICP	MICP Controller	The PC controls each guitar channel input gain per the multichannel configuration while selecting and balancing the gain offsets, the mute state, and the volume inputs from the USB via AICS.
		MICP Device	The guitar gain is adjusted per optimal calibrated configuration, while the PC selects different gain offsets and inputs during the song recording.
	CAP	Initiator	The PC configures the guitar to start audio as Unicast while configuring the guitar source audio locations accordingly; source settings are configured to capture high-quality audio from the room singer and guitarist.
		Acceptor	The guitar accepts each Unicast stream setting and Unicast source setting.
Codec	LC3	Source/Sink	The PC deploys 48 kHz as the source and 48 kHz as the sink to render and capture at highest LC3 quality and at low latency.
Transport	CIG/CIS	CIS Central	The PC captures audio over Connected Streams from the guitar during recording.
		CIS Peripheral	The guitar sends audio over CIS.
	BIG/BIS	Broadcaster	The PC Broadcasts tracks during playback.
		Synchronized Receiver	The headsets receive Broadcast during track playback.
		ACL Central	The PC connects and controls settings such as profile data from BAP, MCP, and MICP, which are sent to the guitar; each cellphone controls each headset to sync to the PC playback Broadcast during the broadcast of a track.
		ACL Peripheral	The guitar receives configuration from the PC; each headset receives Broadcast configuration from each cellphone during playback.

ONLINE GAMING

The next day, you get a recommendation to join a new online game. You turn on your LE Audio gaming headset, and it connects to your PC. While wearing your headset, you hear the game media. The game allows users to start a conversation over a feature called "walki-talki-online." You invite a few friends for a chat via voice commands, and you begin a conversation while playing and listening to the game media. Your hands are focusing on the gaming controls, and each chat session is beginning and ending via voice commands, which allows better focus on the gaming scene, without having to change hand position. While playing, your LE Audio headset plays a notification ring from your front door Broadcaster device. Your pizza delivery just arrived at your doorstep. You ask an online friend to cover for you at the game while you go get your pizza.

Table 8.7 describes how the home recording studio use case is realized using the LE Audio One Architecture.

TABLE 8.7

Home Recording Studio Use Case Layers, Components, and Roles

Layer	Module	Role	Description
App	GMAP	UGG	The PC deploys a Unicast media stream while accepting incoming voice call chats and mixes it into the existing media stream as a new context type update metadata BAP operation.
		UGT	The headset receives multimedia as a sink and sends the source from the microphone when a chat call starts.
	TMAP	BMS	The front door Broadcaster announces a ring when the pizza delivery arrives.
		BMR	The headset syncs to the front door Broadcast announcement system.
Control	BAP	Unicast Client	The PC sets a Unicast audio connection to the headset with a bandwidth reserved to support a two-way call mixed with multimedia.
		Unicast Server	The headset publishes its Codec capabilities to support a two-way gaming chat at high quality.
		Broadcast Source	The front door broadcasting begins when a ring is pressed.
		Broadcast Sink	The headset is configured to receive Broadcast from the front door.
		Broadcast Assistant	The PC scans and configures the headset to sync to the doorbell as soon as the pizza delivery arrived and the ring starts the Broadcast.
	CCP	CCP Client/ Voice commands	The headset accepts a voice command from the headset microphone to accept incoming chat voice calls while in a multimedia game with audio or sends voice commands to the PC to activate various controls and switch chat sessions.

(Continued)

TABLE 8.7 *(Continued)*
Home Recording Studio Use Case Layers, Components, and Roles

Layer	Module	Role	Description
		CCP Server/ Voice recognition gateway	The PC notifies the headset about an incoming chat, so the gamer may accept the incoming chat channel while responding via voice commands.
	CAP	Initiator	The PC configures the headset to start audio as Unicast with a bandwidth reserved for a bidirectional stream to carry a voice chat.
		Acceptor	The headset accepts each Unicast stream setting and Unicast source setting.
Codec	LC3	Source/Sink	The PC deploys 48 kHz as the source and 48 kHz as the sink to render and capture at the highest LC3 quality and at low latency.
Transport	CIG/CIS	CIS Central	The PC sends and receives audio over Connected Isochronous Streams to the headset.
		CIS Peripheral	The headset receives audio over CIS and sends audio back to the PC when voice chat starts.
	BIG/BIS	Broadcaster	The front door announcement system broadcasts the ring when the pizza delivery arrives.
		Synchronized Receiver	The headset receives the Broadcast of the front door ring.
		ACL Central	The PC connects and controls settings such as profile data from TMAP, BAP, and CCP, which are sent to each speaker, and configures the headset to sync to the front door announcement system upon pizza ordering.
		ACL Peripheral	The headset receives configuration from the PC.

EPILOGUE

In this book, we reviewed the evolution of audio in Bluetooth® technology as it manifested in LE Audio. We covered the aspects of audio from the application down to the antenna. LE Audio defines the One Architecture for audio in Bluetooth technology to activate and stream power-efficient audio and addresses use cases never possible before over Bluetooth link, such as Broadcast audio sharing and multistream. The basic building blocks, such as enhanced audio scheduling, efficient LC3 Codec, and a modular host stack, allow for innovation and expansion into new use cases. Use cases such as Broadcast TV Audio and multichannel surround sound are emerging. The invention of new use cases becomes possible using the LE Audio LEGO model application interface. The LE Audio LEGO model is scalable for further generalization by new applications. Longer battery life in audio peripherals becomes possible with LE Audio, and more free spectrum enables multistream use cases. The first waves of the LE Audio Specification by the Bluetooth SIG are completed, but the application opportunities over LE Audio are just beginning to emerge in the years to come.

What's next? History taught us that humans are capable of innovating in unpredictable ways. Science, Arts, and technology evolve hand in hand, while technology is enabling the tools, science is defining the next challenge to overcome, and the Arts provide the vision and inspiration. Culture shifts in our societies present demands for newer use cases, which challenge science and technology. Bluetooth audio is a remarkable success story. With LE Audio, and the new technological advances created by the Bluetooth SIG, there are opportunities to innovate with wireless audio, in ways never possible before. The Bluetooth SIG is working on new ideas, driven by study groups and scientific experiments. For example, the Bluetooth SIG is currently defining a new physical layer for Bluetooth LE®, which will enable higher data throughput while increasing the Bluetooth LE packet lengths to hundreds of bytes. Longer packet lengths use the radio spectrum more efficiently, with less overhead by multichannel audio applications. Other enhancements, which are work in progress by the Bluetooth SIG, will allow an adaptive reconfiguration of an Audio Stream and its QoS parameters, based on link quality, without the need to stop and reconfigure the stream.

The evolution of the Bluetooth LE radio and the physical layer will provide a higher-capacity transport. The LE Audio One Architecture may use the higher-capacity transport to enable concurrent audio use cases, reliable and efficient multi-stream, and enhanced Broadcast Audio Streams. The Bluetooth SIG is defining new radio capabilities to evaluate the accurate location and distance of devices which may be used by LE Audio to enable location-based audio use cases, for example, activating audio as a person approaches a Bluetooth technology-based speaker with a Bluetooth-enabled phone or a Bluetooth-enabled watch. As a result of these new developments, what looked impossible a few years ago is now becoming possible.

How wireless audio experience will look in the next 10–50 years is hard to predict. In this book, we tried to open your imagination to the use case opportunities with LE Audio, and what use cases may emerge in the near future, thanks to LE Audio. Technology and science evolve based on a combination of varying aspects, some of them cultural and some of them economical, or other varying influences by societies. What remains constant is the human endeavor to elevate life quality on Earth.

REFERENCES

1. Bluetooth® Core 6.0 or later, https://www.bluetooth.com/specifications/specs/core-specification-6-0/
2. BAP version 1.0.2 or later, https://www.bluetooth.com/specifications/specs/basic-audio-profile-1-0-2/
3. PACS version 1.02 or later, https://www.bluetooth.com/specifications/specs/published-audio-capabilities-service-1-0-2/
4. ASCS version 1.0.1 or later, https://www.bluetooth.com/specifications/specs/audio-stream-control-service-1-0-1/
5. BASS version 1.0 or later, https://www.bluetooth.com/specifications/specs/broadcast-audio-scan-service/

6. VCP version 1.0 or later, https://www.bluetooth.com/specifications/specs/volume-control-profile-1-0/

7. VCS version 1.0.1 or later, https://www.bluetooth.com/specifications/specs/volume-control-service-1-0-1/

8. VOCS version 1.0.1 or later, https://www.bluetooth.com/specifications/specs/volume-offset-control-service-1-0-1/

9. AICS version 1.0 or later, https://www.bluetooth.com/specifications/specs/audio-input-control-service-1-0/

10. MICP version 1.0 or later, https://www.bluetooth.com/specifications/specs/microphone-control-profile-1-0/

11. MICS version 1.0 or later, https://www.bluetooth.com/specifications/specs/microphone-control-service-1-0/

12. CCP version 1.0 or later, https://www.bluetooth.com/specifications/specs/call-control-profile-1-0/

13. TBS version 1.0 or later, https://www.bluetooth.com/specifications/specs/telephone-bearer-service-1-0/

14. MCP version 1.0 or later, https://www.bluetooth.com/specifications/specs/media-control-profile/

15. MCS version 1.0.1 or later, https://www.bluetooth.com/specifications/specs/media-control-service-1-0-1/

16. CSIP version 1.0.1 or later, https://www.bluetooth.com/specifications/specs/coordinated-set-identification-profile-1-0-1/

17. CSIS version 1.0.1 or later, https://www.bluetooth.com/specifications/specs/coordinated-set-identification-service-1-0-1/

18. CAP version 1.0 or later, https://www.bluetooth.com/specifications/specs/common-audio-profile-1-0/

19. CAS version 1.0 or later, https://www.bluetooth.com/specifications/specs/common-audio-service-1-0/

20. LC3 version 1.0.1 or later, https://www.bluetooth.com/specifications/specs/low-complexity-communication-codec-1-0-1/

21. HAP version 1.0.1 or later, https://www.bluetooth.com/specifications/specs/hearing-access-profile-1-0-1/

22. TMAP version 1.0 or later, https://www.bluetooth.com/specifications/specs/telephony-and-media-audio-profile-1-0/

23. GMAP version 1.0 or later, https://www.bluetooth.com/specifications/specs/gaming-audio-profile-1-0/

24. PBP version 1.0.1 or later, https://www.bluetooth.com/specifications/specs/public-broadcast-profile1-0-1/

Index

Note: Page numbers in *italics* indicate a figure and page numbers in **bold** indicate a table on the corresponding page.

For Product Safety Concerns and Information please contact our EU
representative GPSR@taylorandfrancis.com
Taylor & Francis Verlag GmbH, Kaufingerstraße 24, 80331 München, Germany

www.ingramcontent.com/pod-product-compliance
Lightning Source LLC
Chambersburg PA
CBHW031954180326
41458CB00006B/1710

* 9 7 8 1 0 3 2 9 6 6 1 9 9 *